Your ( ) Unzipped

Your Genes Unzipped

How Your Genetic Inheritance Shapes Your Life

Tim Spector

ROBSON BOOKS

First published in Great Britain in 2003 by Robson Books, The Chrysalis Building, Bramley Road, London, W10 6SP

An imprint of **Chrysalis** Books Group

British Library Cataloguing in Publication Data
A catalogue record for this title is available from the British Library.

ISBN 1 86105 662 1

Typeset by SX Composing DTP
Printed by Creative Print & Design (Wales), Ebbw Vale

Contents

Acknowledgments   ix
Preface   xi

Chapter 1: Who Are We?   1
Isn't he marvellous!   1
Loving our genes and diseases   3
A (very) short history of evolution   4
Where do humans come from?   5
What are genes?   7
Who are we?   12
Nature or nurture: how can we tell if a trait
    or disease has a genetic basis?   15
Gene hunting   18
Genetic testing, now and the future   19
Are genes all that matter?   21

Chapter 2: Genes, Worry and the Early Years   24
What will my baby look like?   24
Will our next child be a boy or a girl?   28
He has the milkman's eyes   30
What's in a name?   32
Cot death   35
Parenting instincts   37
Why incest is not best   40
'Designer babies'   43

Chapter 3: The Early Years 47
Sugar and spice 47
Gameboy eyes 49
How tall will he be? Greens or genes? 52
The bed-wetter 54
Wheezes and dirt 55
Milk and good allergies 57
The difficult kid 59
The geek gene 61
The slow reader 63
Are there genes for intelligence? 66

Chapter 4: Genes and the Terrible Teens 70
Why is life so unfair? 70
Budding too early? 72
Spots and chocolate 74
Early habits 76
Skin and bone 79
Off-key genes 81
Lazy genes 83
Champion genes 85
My son the bully 90
Wallflower genes 92

Chapter 5: Genes, Attraction and Sex 95
Fatal attraction 95
Do genes smell attractive? 98
What do men want? 101
The one-night stand 103
Sexually incompatible 104
Orgasm and the G spot 109
Male infidelity 112
Female flings 114
Jealousy and competitiveness 119

Cads, dads, sex and food	121
Male problems: all in the jeans?	124
Love genes	127
Is a successful marriage genetic?	129
Gay genes	132
AC/DC	136
A man's best friend?	139
Chapter 6: Grown-up Genes, Instincts and Risks	141
Can't give up?	141
Drunk in charge of your genes	145
Living on the edge	147
Greed and happiness	150
A matter of taste	153
Junk-food junkies	155
Occupational hazards	158
Fears and phobias	160
The sucker gene	163
Heroes and cowards	165
Chapter 7: Minor Body Irritations	168
A pound of flesh	168
My bum's too big	173
Chilly fingers	175
Bugs and bites	177
Travel sick	179
Early wrinkles	180
Colds, sniffles and HIV	183
It's on the tip of my tongue	187
Sleepless nights	189
Medicines don't agree with me	192
Baldness and macho genes	195
Toothless	197
PMT genes	200

Flying risks   202

Chapter 8: Genes, Diseases and Getting Older   205
Will I get my parents' cancer?   205
Breast-cancer risk   207
Moles, sun and skin   209
Ageing and longevity   212
Biological clocks and genetic grannies   216
Brittle bones   218
Family back problems   220
Self-destruction   222
Relatives behaving oddly   225
Losing your mind   228
Infectious immunity   231
Dropping dead   233
Will I die young of heart disease?   235
Why do females usually outlast males?   238
Lumps and bumps   240

Chapter 9:   Beliefs, Morals and the Afterlife   242
Political leanings   242
Criminal behaviour   244
The darker side of the Y chromosome   246
Skin deep   250
Genes for God?   255

Chapter 10: Conclusion   259
Redressing the balance (ironing out your genes)   259
Variety is the spice of life   263
Playing out your hand   266

Glossary   268
References   271

Acknowledgments

Few of the ideas in this book are novel or original – most have been 'synthesised' from the works of the many contributors to this field, both well known and more obscure. Useful further reading and key references for each chapter are given at the end of the book. Any mistakes and erroneous theories, which time may disprove, are entirely my own. Many people have helped with this book at some level – from Bryan Sykes of Oxford University, who first encouraged me to study twins and enter the field of genetics, to friends such as Neil and Brenda Jordan and various skiing companions, and the author Pam Spurr, who prodded me to write the book. Credit goes also to friends and colleagues who diligently read the early drafts, such as Mike Fitzgerald, Rita Euziere, Mitch Blair, Lawrence Rubin, Lesley Bookbinder, Penny Wright and Francine Hochberg. In particular, I need to thank Lynn Cherkas, Susan Hochberg and Deborah Phippard, who dedicated large amounts of their time and made many improvements. I'm grateful to Pat and Victoria for secretarial help. Thanks also to Jo Brooks and her team at Chrysalis Books for essential help in editing, and for her encouragement. I'm also grateful to Nick Martin and David Lykken for providing helpful extra genetic data, to Lynn Cherkas for performing additional genetic analyses on our twin dataset and to John Blangero for his helpful personal insights into human and baboon sexual behaviour. The Chronic Disease Research Foundation (CDRF), a small charity, supports much of our twin research. I'm grateful to my fellow workers at St Thomas' and the Twin Research Unit as well as worldwide genetic colleagues and my European colleagues from the GenomEUtwin research group and many others for continuous snippets of information. Lastly, I need to thank my family and friends for providing the ideal environment, and my parents and the twin volunteers for providing the essential genes.

Preface

There are many excellent books on genetics, so why write another? Talking to friends, colleagues and patients over the last two years convinced me that there was a major need for a resource for the nonscientific reader. Most of the books currently available tend to focus on distinct areas and are either very modern and technical in approach or directed towards understanding how our species – called humans – have developed and evolved over the last few million years; few are both. Some books try to explain our fascinating natural instincts in terms of some general animal model, without explaining why as individuals we differ so much from each other.

The unique format of this book – encompassing human life from its beginnings to its end – allows us to explore the genetic components of the common ailments and diseases. The role of genes in common diseases is usually overlooked, as are our risks of getting them. This is because scientists initially (and wrongly) believed that the main influence of genes was on rare, fatal diseases or cancer. Most medical specialists find the field of genetics to be moving so fast that few can keep up sufficiently to provide advice across a range of problem areas. My work over the last twelve years using the UK's largest twin registry has, fortunately, opened up the prospects for exploring all aspects of human life, health and disease – crossing traditional professional barriers of specialisation in the process.

This book is aimed at everyone who wants to learn more about how genes affect their everyday lives and the lives of their relatives. It should encourage you to find out about your family history – in terms of diseases, personalities and life events – allowing you to gain insight from the experiences of your parents, brothers, sisters, grandparents and even your ancient ancestors. The knowledge you gain about what makes you and maybe your partner tick can alert you to your desires, your likely strengths and weaknesses and what problems to avoid in later life. With this insight you can maximise your environment and surroundings to beneficial effect. Genes have been passed down to you as a consequence of endless random replication over the generations, during which particular combinations of genes prospered because they had evolutionary benefits, such as aiding survival or helping you to find a mate and reproduce. Even if the gene seems a bad one to you, it usually confers some advantages in other aspects of your life or assisted your ancestors' lives – or it wouldn't be there. Sometimes just knowing that a problem you have is inbuilt, and part of a genetic package deal, can be helpful in the way you deal with it.

Your Genes Unzipped aims to be accessible to the layperson. I have tried to use as little jargon as possible and to reduce the scientific detail to the essentials. Hopefully, this book will be both enjoyable and informative. This simplified black or white view of the facts will undoubtedly annoy more academically minded readers, to whom I apologise in advance and try to placate with an up-to-date reference list for further reading.

Much of the data presented comes from the St Thomas' UK adult twin registry, which is based on nine thousand twins, and, being new, has not yet been published in full. The scenarios used in this book are a way of recasting genes into the context of everyday situations that many of us will encounter at some time in our lives. Some of the stories are based on actual case histories, while others are composites. In the scenarios I have used initials rather than names. This serves two purposes: first, to preserve anonymity, and, second, to allow you to imagine the scenario within your own cultural and environmental context, helping you identify more with the characters and situations whatever your age, health, gender or nationality, whether the heroes, villains and by-standers be your parents, children, friends or even yourselves.

Unfortunately, you can't adequately address all of life's many problems, emotions and diseases or their solutions in a short book. The examples here should hopefully help you cope with or better understand similar problems or situations – but are no substitute for seeking professional help or assistance.

Tim Spector
Professor of Genetic Epidemiology,
St Thomas' Hospital, London

1

Who Are We?

As a prelude to a closer understanding of the key question of who we are, we start with a simple introductory scenario. This is followed by an explanation of a few fundamentals in genetics that are useful in order to get the most out of this book. For those of you new to the area, try not to be put off by the jargon – there are only about ten words you really need to understand and there's a glossary at the end if you forget.

Isn't he marvellous!

On the day J graduated with a first-class degree in science at the top academic university in the country, he already had many prestigious job offers to contemplate. Most of his family, including his parents, two aunts, two grandparents and younger sister, were there to congratulate him. J was the first member of his family – at least that anyone could remember – to have excelled at school and attended university. Everyone congratulated the parents – who ran the family bakery – on how they had brought up J to be so intelligent, hard working, considerate and polite. His mother felt very proud. However, everyone felt a bit sorry for his older brother, who had left school early (after truancy and many problems) and was finding it hard to get a job, and his younger sister, who was a struggling hairdresser's assistant. Why had the three children turned out so differently? What had made J so special? He was shorter and less muscular than his brother – but was that just because he did less sport, or was there another reason?

Meanwhile, 7,000 miles away, someone else's 21-year-old son, M, was in deep trouble. His parents had just been allowed into a Malaysian prison where he was facing a possible death sentence for possession of drugs, having been arrested at the airport. His parents were in despair, asking themselves, 'What did we do wrong?' He'd gone to the best private schools, had personal coaches and tuition – but, despite these additional resources, he had always seemed restless and out of place and always taken risks: had got into drugs early and had been in constant trouble with authority. From birth he had acted differently from his brother, who, in total contrast, was a model son. Doctors had told them that as well as having diabetes, M had a personality disorder, but neither parent knew anyone with either condition in their families. Naturally they felt guilty and believed that in some way they were responsible.

Both J and M had a secret in common – unknown to either of them. Both were unrelated to the people they believed to be their mother and father. They were two of the estimated one in ten to twenty thousand babies in the 1960s and 1970s who were mistakenly switched at birth. Blood tests confirmed that these errors did happen and were often due to obsessions with hygiene and large common rooms with incubators. These mistakes are thankfully much less common nowadays, as mothers are encouraged to bond with their babies earlier. The scenarios of J and M give a taste of the importance that our genetic makeup has on all aspects of our lives.

When parents are proud of their offspring, they usually take the credit for their parenting abilities as well as the skills they have passed on. What happens when children don't succeed at school or socially? It is common to attribute blame to society or someone else. But whose fault is it? What if most of our actions today were programmed by evolutionary events involving sex and survival maybe millions of years ago?

Loving our genes and diseases

Most of us fall in (and out of) love at some point in our lives. But what is love? Are we programmed by our parents, friends or the media to be attracted to a certain type of person? Is it totally random or something we learn? What if we are just subconsciously following the programmed (and perhaps flawed) tastes of our distant ancestors? Men and women over the course of millennia have developed increasingly different roles and different tastes, attitudes and mechanisms to benefit their own sex's survival. Could the habits and instincts of your ancestors really alter the way you behave towards the opposite sex in the twenty-first century?

All of us will inevitably contract diseases in our lifetime. Predicting those we are most likely to suffer from would allow us to take preventive action. For the last thirty years, experts have told us that our problems are self-induced – because of our poor diets or lifestyle. They tell us that, if you eat lots of raw fruit and vegetables and live like a saint, all of us will live to be a hundred. Unfortunately, life isn't like that, and the role of lifestyle and diet in diseases has been grossly exaggerated. Many people's lives have been made a misery by faddish, short-term, impractical and ultimately useless diets – and others have become very rich. Diseases are due to our ancient natural defence mechanisms either breaking down under stress or overreacting to our environment. We all inherit these defences from our ancestors through the transfer of genes. Many modern diseases are due to a crucial interaction between our ancient genes and recent changes to our environment or lifestyle – which they weren't designed to cope with. A good example we shall explore later is being fat: genes that allowed you to maintain weight in times of hardship were a great advantage, but, now we are surrounded by too much food, they can be a major disadvantage. If you know at an early age that you are predisposed to obesity, avoiding certain foods and knowing you need to burn more calories could prevent the weight gain that may change your body's metabolism for ever. Overall, your family history is an infinitely better indicator of your future risks of disease than how many carrots or vitamins you eat per week.

A (very) short history of evolution

Charles Darwin's theory of evolution by natural selection, written in 1859, has stood the test of time as one of the most important discoveries ever made. After studying animals on isolated islands, Darwin concluded that all animals then in existence had evolved from other animals and species since life began some 4 billion years ago. He believed that in each generation there were tiny random alterations in physical form. If these alterations improved survival or reproduction it was preferentially passed on to the next generation, until most of the population had the trait. What he couldn't figure out was how this information (heredity) got passed down to the next generation. Around the same time, a Czech monk, Gregor Mendel, while working on peas in his garden, discovered how simple traits such as size and colour were transmitted from parents to offspring in a simple binary (1 or 0) system – with equal shares from each parent. This introduced for the first time the concept of the passing on of separate packages of information about traits down the generations, via unique entities: called genes.

With the twin concepts of slow natural selection (evolution) and an understanding of how the information is transferred (genetics), life on earth had suddenly become much easier to explain.

For the first 6 billion years or so since the universe began, nothing much appears to have happened. Then, by some accidental combination of chemicals, life started in its simplest form: an entity that could replicate itself, surviving in water or in hot sulphurous rock and receiving energy from the primeval soup around it. During the copying process, a few 'typos' occurred at random, some of which provided benefits for the next generation in the environment in which they lived, while others were disadvantageous. To protect themselves, the life forms evolved protective shells, becoming early types of bacteria and viruses. More complex single-celled entities evolved with specialised energy sources. Some of these grouped together into multicelled cooperative colonies, each with its own set of identical replicators (genes). These multicelled creatures became gradually more complex, each cell starting to have different functions, but all having the same genes. Eventually, simple fishlike creatures developed. These became gradually more sophisticated, and some had mutations that allowed them to live on land as small reptiles around 300 million years ago, and mammals with warm blood (such as apes and humans) developed from them.

In different species, climates and habitats, the useful genes survived and the bad ones didn't. Evolution is a mechanism for living things to slowly keep pace with the constant changes in the surrounding environment permitting survival and replication. Not all living creatures are evolving. Some, such as the ancient fish called the coelocanth, stopped evolving 300 million years ago, having become perfectly adapted to the planet's environment. Humans are still evolving, as our environment and the life forms around us (such as bacteria and viruses) are constantly evolving in ways that threaten our survival.

Where do humans come from?

Humans started to separate from our ape ancestors fairly recently (only about 5 million years ago) in Africa, developing both an ability to walk on two legs and a superior brain function. Initially, there were several different types and species of ancient (archaic) human, which may have co-existed. *Australopithecus* was one of the first we know of, and then our own genus *Homo* emerged with several distinct groups. These included *Homo ergaster* and *Homo heidelbergensis*, who didn't make it very far. *Homo erectus* did manage to get out of Africa and had spread to Indonesia by 1 million BC – but for unknown reasons became extinct.

Modern man comes from the group *Homo sapiens*, which emerged in East Africa around 150,000 years ago. This group migrated from Africa to Asia and became recognisable as modern humans only about fifty to a hundred thousand years ago. They are sometimes known as 'Cro-Magnons' after the caves in France where skeletons were first found. Despite the fact that Darwin's theory fits all known scientific discoveries and fossil and skeleton collections, many people around the world still don't believe in evolution. A Gallup survey in 1997 of 12,000 Americans found that 44 per cent disbelieved the theory and, despite all the evidence to the contrary, thought the world was less than ten thousand years old. This compares with a mere 7 per cent of Britons who hold similar views. Only 10 per cent of Americans believe that man evolved from simple life forms without God's help – although among US scientists this increases to 55 per cent. Many people from a variety of religious backgrounds believe in an original human couple, Adam and Eve, created as the Bible describes in the Garden of Eden only 4000 years BC. Despite this, most people are prepared to believe in genes, cloning, gene therapy and the prospect of new wonder drugs to combat all our ills.

We now know that most of the evolutionary changes to our genes occurred in a (relatively!) short time period in history, in the latter half of the so called Pleistocene period, which was between 1.6 million and 10,000 years ago. At that time of great change around 100,000 years ago, our modern human ancestors were migrating out of equatorial Africa into the Middle East, crossing into Asia and Europe as small travelling bands of around thirty people, surviving the last ice age, crossing the dried-up English Channel and other seas. While doing this they were learning to speak, using better tools and weapons, evolving larger brains, busily mating with one another and producing children – they were, unbeknown to them slowly selecting the genes that we have today. As if this weren't hard enough, they had to stay with the same group of people most of the time. Imagine a whole lifetime with only your family and in-laws for company. Most people find a weekend hard enough – so our ancestors must have been tough!

Modern humans (*H sapiens*) displaced earlier models of humans such as the tough, sturdy and (contrary to stereotypes) large-brained Neanderthals (named after a valley in Germany where bones were found), who had been in Europe for over 100,000 years. They never evolved much and eventually died out 30,000 years ago. We don't know whether modern man mixed or mated with Neanderthals, as we have no relics of their genes in us today. The evidence to date suggests the two groups stayed apart. The pressures and priorities for early modern humans were to survive in harsh conditions and produce children. Those that were good at reproducing gave us our modern genes; those that failed are fossils, bones and ancient history. The legacies they gave us shape our individual desires, skills and – unfortunately – diseases.

What are genes?

Genes are little natural self-replicating machines that provide the blueprints for building proteins. Proteins are what we are all made of. They make, start and drive all our body's structure and chemical reactions. Genes are useless by themselves and the proteins they produce do all the work. These protein reactions are responsible for the way we think, talk, act, look, eat and breathe – as well as our subconscious desires. Genes are the blueprints for life. We all have around thirty thousand of them contained in every single one of the 100 trillion (or so) cells in our body. These genes are spread out along 46 chromosomes, which are wavy strands bundled together in 23 pairs in the centre of every cell. When a cell divides it makes a copy of itself and duplicates the chromosomes and therefore the genetic blueprints.

Humans all have the same basic menu of thirty thousand genes – but the forms of each gene vary between us in subtle ways. Our genes are composed of the substance DNA

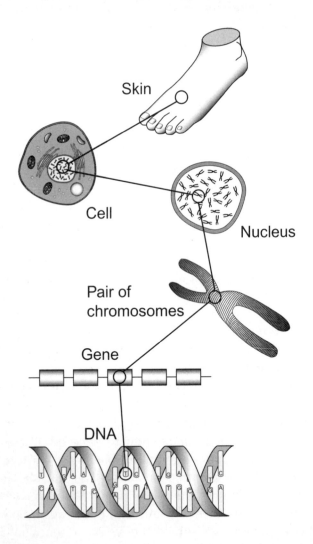

Skin

Cell

Nucleus

Pair of chromosomes

Gene

DNA

(deoxyribonucleic acid). A complete list of all our genes is arranged as a string of 3 billion chemical building blocks called base pairs. Written out in full, this simple genetic code (genome) found in every cell of our body would fill an average-size book, 600,000 pages long. DNA is arranged in a clever spiral ladder formation whose structure was discovered in 1953 by Francis Crick and James Watson in a lab in Cambridge (and the Eagle pub nearby). The secret to DNA's ability to copy itself lies in its unique double-stranded structure, which can unravel. Each single strand is then copied, producing two identical versions with the exact same sequence of bases to pass on to the next generation.

We inherit 100 per cent of our genes from a combination of those of our parents. We receive half our genes from our mother and half from our father. The differences between us are a result of inheriting different *forms* of the same gene, rather than inheriting totally different genes. For simplicity, here we refer to genes rather than forms of genes. Imagine a game of cards, where your parents each have two packs. To reproduce, your parents produced eggs or sperm, which each contained only one pack of cards, made up by randomly shuffling their two original packs, then halving them and throwing the other half away. All sperm and eggs contain different combinations of genes, so no two sperm or eggs will be genetically the same. When egg meets sperm, the father's and mother's pack are shuffled together to create two unique packs, and therefore a new individual with a unique combination of parental genes. The only exception to this rule is the case of identical twins, which occur when the fertilised egg copies itself very early on and each twin gets two identical packs.

We share on average 50 per cent of our genes with each of our parents, children and brothers and sisters – but the individual genes shared differ for every relative. Using the card analogy, you may share on average half your cards with your brother and mother, but the queen of spades you hold is shared only with your mother, and the jack of hearts your brother holds is not in your hand. You share 25 per cent of your genes with all your grandparents, the same proportion as with your blood uncles and aunts. The genetic sharing decreases between first cousins or between you and your

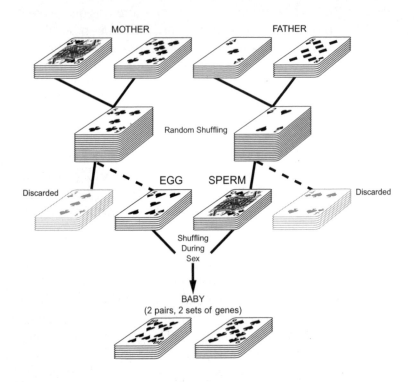

The card/gene analogy

great-aunts and great-uncles at only 12.5 per cent, and drops to only about 3 per cent with your second cousins.

Mammals use sex unwittingly to mix the genes passed on to the next generation. This mixing has a number of advantages. As we inherit 23 pairs of chromosomes, one from each parent, any bad genes on one of the pair can be neutralised by normal genes in the other pair. This keeps the number of effective bad genes to a minimum – unless you're unlucky enough to inherit the same rare bad genes from both parents and therefore have two copies.

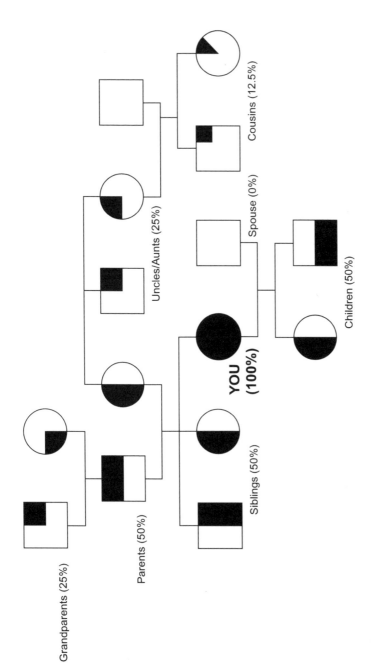

Proportion of your genes shared with other family members

By continually changing gene combinations, nature ensures that all children won't have the same reactions to the environment or parasites, and therefore the same diseases, and die out. The mixture may sometimes be fortuitous and produce a novel variation of a gene useful for survival or reproduction. The owners of the best and most diverse combinations will generally do the best in evolutionary terms when, as in man, there is a constant running battle with the environment and each other.

Overall, our gene sequences are 98.5 per cent identical to those of chimpanzees, with whom we shared common ancestors 5 million years ago, the so-called 'missing links'. We are genetically closer to chimps than chimps are to gorillas – making some scientists propose we are the same species. Our genetic blueprints are so common that we share the same bones, body chemicals, immune responses and even brain structure. The main differences between us and chimps are not the total of number of genes, but the number turned on and activated (expressed) in our brains. Chimps have the same genes but many are dormant, and for some reason were never switched on. We think ourselves to be pretty smart – but our genes also have a lot in common with very simple animals such as the worm, who share 50 per cent of our genes, or the 'sophisticated' banana with which we share 30 per cent of genes. This demonstrates that it is not the *number* of genes that gives us our complexity, but the *way they interact* with each other and our environment. In evolutionary terms, our genes are likely to be virtually identical to our early human ancestors of only a hundred thousand years ago.

Who are we?

When people ask you, 'Where do you come from?' it's getting harder to answer. Is it the place you now live in or where you were born? Or is it where your parents or grandparents were born? Try going back only four generations to your 32 great-great-grandparents, and you will see how difficult it is to classify yourself. Ten generations ago we had 1,024 great-great-great-great-great-great-great-great-great-grandparents. Some people try to define themselves ethnically, citing religious or cultural groupings such as Jews and Muslims, but these are also relatively modern, going back only a mere 120 or 200 generations. Many other older and once famous ethnic groups such as the Sumerians, Etruscans and Philistines have long since disappeared. The only clear answer to where you come from is that 7,500 generations ago we were all East Africans.

Language can also reveal our genes and origins. Most European languages are of the Indo-European type, which originated in the Middle East. Our farming ancestors who took over from the original hunting inhabitants used it. They and their ideas migrated slowly outwards from the Middle East, the so-called Fertile Crescent, where agriculture began some thirteen thousand years ago. These people spread with their domesticated cattle and sheep, wheat and barley slowly northwards and westwards, bringing their ideas, language and technologies with them at a rate of half a mile a year. By 5000 BC farming had reached southern Spain and by 3800 BC had reached southern Britain and Scandinavia. The original occupants took up farming and the cultures that went with it, or died out or moved to the extremes of Europe.

So the distance from the original farmers based around the River Jordan is a marker of how each population is related. Therefore, Britons are more closely related to Spaniards and Portuguese than to Greeks or Yugoslavs. Some peoples such as the Basques and the Finns, who have unique languages, remained genetically distinct, probably because they stayed as hunting populations and resisted farming till much later. Most northern Europeans today have 25 per cent of their genes clearly identifiable as of Middle Eastern origin.

We may, however, not be the distinct groups or races we think we are. Men and women each carry a very special subtype of DNA, which acts like a genetic fossil record of our past. These special bits of DNA (the Y chromosome in males and mitochondrial DNA in females) are carried only by the egg or the sperm, and so don't get mixed in the normal way. These two special types of DNA, unlike the rest, have therefore changed relatively little over the generations and give amazing insights into our past and origins. By looking at the patterns and similarities of your Y chromosomes, you can trace and compare all your male ancestors and by looking at mitochondrial DNA you can work out your links with other females or your female ancestors.

With the exception of a few areas on the north coast that have Viking genes, Scots are no different genetically from their age-old opponents the English, whereas the Irish and the Welsh are more distinct. The English are in fact a complex mixture of Bronze Age Beaker folk, Iron Age Celts, Angles, Saxons, Normans, Romans and Middle Eastern farmers. Africans from Ethiopia are quite distinct from the Bantu further south. Surveys have shown that people considering them-selves to be black African Americans or black Afro-Caribbean in the UK all have some white ancestors and have genes that are on average 30–40 per cent European. Similarly, many people who think of themselves as white, particularly in the US, have recent African genes.

One of the more prominent examples of this is the debate as to whether Thomas Jefferson, the third US president, had produced a love child with his fourteen-year-old slave, Sally Hemmings. The genes, and in particular those on the Y chromosome of the descendents of the Hemmings and Jefferson families, were recently compared, and the markers found to be identical. Although it is possible his brother Randolf had also been tempted, Thomas had no alibi for the nine months preceding the pregnancy, and is the likely father. US presidents nowadays need to be more circumspect where they leave their genes and bodily secretions.

Nature or nurture: how can we tell if a trait or disease has a genetic basis?

For doctors and health workers, knowing whether diseases and traits are predominantly genetic or environmental is crucial. There are a number of methods that vary in accuracy or usefulness depending on the question being asked. For instance, you can ask a group of people with the disease or trait (we call these people the *cases*) whether their parents or relatives also had it and then compare the family history with the responses of normal people without the disease (*controls*). If the individuals with the disease or trait have a greater family history of it than normal people, this supports evidence of a genetic influence and is called an increased risk. However, these studies can often be biased. People with disease may (perhaps unconsciously) exaggerate the numbers of affected relatives and normal people will often underestimate them.

You can also look within families to see if close relatives (i.e. first-degree, sharing half the genes, such as brothers, sisters and parents) are more similar to each other with respect to the trait than more distant relatives who share fewer of their genes (e.g. cousins). If close relatives are more similar, this is called *familiality* or *family clustering*. Both these designs provide a good first clue to genetic factors, but don't separate out the effects of genes from the effect of shared family lifestyle or upbringing, because family members also share these. For many traits and diseases, results can be confused because of age differences between the generations

Twin studies are the ideal natural experiment to separate out the effects of nature (genes) and nurture (environment). Crucial to this is the fact that there are two biological types of twin. Essentially, twin studies compare the similarities of a group of identical twins (sharing all their genes) with a group of nonidentical twins (sharing half their genes – like brothers and sisters). Both sets share the same environment (in the womb and childhood) and so any difference between them must be as a consequence of genetic factors. Note that twin studies can't be used for very rare diseases or where the trait of interest is related to *being* a twin (e.g. being a small baby). Also, they can tell us something about a trait only if there is actually some genetic variation (difference) between people. For example, twin studies wouldn't tell you anything about the ability of adults to eat, as all humans have this skill. Twin studies can tell you how much of the difference (variability) in a trait between people is due to genes and how much is due to environment. This proportion is called the *heritability* and this term is used throughout this book, often as a percentage. Some people suggest twin studies overestimate genetic influence for some traits, since parents might in theory treat identical twins differently – but there is no evidence in real life that this is an important factor.

The first ever twin study was performed in 1924, when numbers of moles and freckles were counted and found to match much more closely in identical twins than non-identical (fraternal) twins. The study concluded that moles and freckles were genetic and not caused by the birth itself. Around one in eighty live births is now a twin, two-thirds being fraternal. Rates in most countries are increasing rapidly due to better survival rates and the increasing use of fertility treatments. In many countries one in four newborn twins is conceived by in vitro fertilisation (IVF), and this figure is set to continue increasing.

Adoption studies are also used as an alternative or addition to twin studies to answer the nature-versus-nurture issue. Siblings or twins who are separated at or soon after birth and raised in different families are studied. The adopted children can be compared with their twin or sibling as well as with any unrelated family members they were raised with. This is a powerful method of distinguishing the effects of genes and the environment, but, as separate adoption of children from the same family becomes less common, the size of the studies is often small.

Breeding experiments – usually using laboratory-reared animals such as rats, mice, worms or fruit flies – are used to test how traits such as behaviours or hair colour are transmitted. Once a gene has been identified in humans it can be introduced into the DNA of animals such as mice to see which proteins are made and how they respond in order to try to understand how the gene works. This has led to many major science breakthroughs, although unfortunately not all genes behave the same way in humans as in other animals.

Studying animal behaviour, as Charles Darwin did, by observing similar species in different environments tells us which traits and characteristics are likely to have developed slowly by evolution to suit new situations – a selective adaptation as a result of gradual changes in the genes over generations. Comparing the behaviour of different groups or species who used to be closely related but were separated from each other at different times also helps us to understand how genes changed in response to different environments. An example is comparing the sexual behaviour of gorillas with that of chimpanzees and humans, or the changing physical shape of Darwin's finches on the Galapagos Islands.

Finally, studying preferences, habits and desires in men and women across different races and cultures – ranging from Amazonian tribes to Californian students – can provide useful insights. If there are consistent findings in many different societies, these are unlikely to be due solely to cultural pressures (e.g. advertising or religion) and more likely to be evolutionary and a consequence of genetic forces. These studies have to be very large and wide-ranging to be useful, and interpretation is always speculative and should be viewed cautiously. For example, studies like this may find that all humans in the world sleep lying down or throw spears with the sharp end first, but this doesn't mean that there are particular gene variations for these traits.

Observations of the patterns of life of a few small hunter-gatherer populations may also provide clues as to the likely lifestyle of our ancestors – but these populations have by now been 'spoiled' by contact with modern man and many intrepid but curious researchers. We have only the early accounts to go on. Critics of this method say that nearly any modern behaviour can be justified by any imaginative evolutionary theory.

When considering a question about behavioural traits or disease, it is most effective to obtain evidence from several different sources. In this book I commonly provide evidence from modern twin studies to prove that a trait is heritable, supporting previous evolutionary theories and backed up where possible with evidence of the involvement of a particular gene.

Gene hunting

Having found evidence of a genetic influence on a trait or disease, the final proof is to find the gene or genes responsible and work out their exact function, the proteins they produce and the effect of small variations or mutations in the gene. This is often a lengthy process. Rare diseases (such as cystic fibrosis or haemophilia), usually affecting fewer than one in ten thousand people, are caused by a problem in a single gene – one or two families were usually enough to pinpoint and find the gene. Several thousand single-gene diseases have now been identified.

For the common traits and diseases discussed in this book (so-called complex traits), finding the relevant genes is much more difficult. More than a hundred genes may be involved, each contributing only a tiny effect. For these complex traits, there are two main methods: studying families and studying unrelated individuals. Both methods require hundreds or thousands of subjects. One approach in the family design is to study several large extended families with hundreds of related people over several generations, who are then examined for the trait or disease of interest and the results compared with markers on their genes. Alternatively, large numbers of brother–brother or sister–sister pairs can be used where one or both have the disease. Only a few clear examples of genes found in this way have so far been reported – mainly because scientists have underestimated the numbers of subjects needed for such studies.

The other way of locating the genes that influence a trait or disease is to compare the DNA of large groups of people who have a particular disease or trait (cases) and normal subjects (controls) for differences in thousands of gene markers. Any differences found could be due to chance and so the findings need to be replicated many times in different populations before it can be ascertained that a particular gene is involved – and this is hard work.

Many diseases and traits are particularly difficult to study – in the same way as personality is difficult to define by questionnaires. In these cases, designing specific experiments where the environment can be closely controlled is proving to be more productive. Once a gene is found it has two uses: first, it might make a diagnostic test for a disease or trait; second, it can tell us enough about the biology to make a new pharmaceutical drug from the newly discovered pathway.

Genetic testing, now and the future

The human genome has now been 99 per cent sequenced and its 3 billion characters are being meticulously translated. You can now buy a copy of your own genome on a CD-ROM in the USA for around $500,000 – although it'll cost you a lot more to understand it. If you bought the CD-ROM expecting the secret of life, you might be disappointed, since only 2 per cent of it codes for genes. The rest is evolutionary baggage, comprising endless codes of gibberish, old genes that have stopped working, sections of DNA from viruses that tagged along for a free ride (and replication) and many other bits that we don't know the function of. However, some of this so-called 'junk DNA' may turn out to have a key role in the fine tuning of our genes – making us more complicated beings than our mere thirty thousand genes would make us think.

Genetic testing goes on already in most developed countries. Routine screening is performed for a number of genetic diseases at birth and all include tests for thyroid disorders and a condition called phenylketonuria. Both are fairly rare single-gene (Mendelian) disorders but curable by early treatment or change in diet. It is likely that, in ten years, genetic tests for many of the common diseases such as heart disease and dementia will also be in routine usage, and ten years after that genetic tests for all the common personality traits will be possible – if we want them. There are a number of genetic tests in existence today for a few rare diseases, such as familial breast cancer and Huntington's chorea, a disease causing early dementia.

Currently, the two most common routine genetic tests in use today that potentially affect us all are DNA fingerprinting, used to link criminals with the scene of the crime, and DNA family testing to prove biological relationships such as who is the real father (paternity testing). Both these tests have had major effects on law and are starting to influence society. Recent studies of DNA have been used to identify whether the charred remains of a family found buried in a wood in Russia were the last of the Romanovs, the tsar's family shot in a cellar at the beginning of the Russian Revolution in July 1918 (they were). DNA was also used to determine whether the body of Joseph Mengele (the Nazi concentration-camp doctor) had been found (it had).

In both cases, scientists confirmed the likely identities by comparing the DNA from the remains to living descendents. In the case of the Romanovs, a possible link was traced by comparing the mitochondrial DNA found on the bodies to that inherited by Prince Philip, Duke of Edinburgh, via his mother's mothers, who were Romanovs. More recently, DNA testing was used to identify the fragments of victims of the September 11th World Trade Center attack and to assess whether Saddam Hussein was dead or alive.

The first person to be convicted on DNA evidence was Colin Pitchfork in 1987, who had raped and murdered a schoolgirl in 1983 (a man who had previously confessed was released after tests on his DNA showed he could not have been responsible). Since then, in the US, 124 convicted criminals (including 37 for murder) have been released thanks to DNA evidence. Recently a law was passed in the UK that allows the DNA of all past and present criminal suspects to be kept on a central database, so that genetic matching by the police will become as routine as fingerprinting.

Paternity testing is now becoming widespread even outside showbiz and Hollywood circles. One of the first cases concerned Charlie Chaplin and his alleged child. The crude but fairly reliable ABO blood-group system was used. Chaplin was found to be genetically innocent, but paradoxically still found guilty by a US jury. Nowadays paternity testing is a fairly simple procedure advertised on the Internet, where hundreds of companies offer services. For around two hundred dollars, it involves rubbing a stick around the lining of the mouth of whoever is being tested, and sending it to a lab, where seven to ten DNA markers are examined within a few days. The manufacturers claim error rates of less than one in 100,000 – but if correctly carried out (without contamination), because of the immense diversity between individuals, the errors should be less than in one in a few billion.

At the moment these tests are not well controlled or supervised and the results can be difficult to interpret and very traumatic for all involved – often it's best not to know the result. As science recognises the overwhelming importance (both good and bad) of genes in our lives, innumerable tests like these could appear in the future – with potential benefits and pitfalls.

Are genes all that matter?

Life would be far too dull and predictable if genes were all that mattered. All traits, diseases and behaviours in humans are to varying extents a mixture of genes and environment. No traits or diseases are purely genetic or purely environmental. We are all warned that cigarettes ('environment') cause lung cancer, but if every adult on the planet smoked to the same extent, only 10–15 per cent of people would get lung cancer. Those contracting cancer would do so because of their genes – and lung cancer would be seen in that scenario as a 'genetic' disease. Currently, in countries where around 25 per cent of adults smoke, it is seen as a purely environmental disease – as those who get cancer without smoking are very rare. Genes don't operate in isolation. They work or are 'turned on' if other key elements are present: other genes, for example, or when the body reaches a certain age, or only in specific environments (such as cigarette smoke). Similarly, genes can be turned off – as happens for many growth genes when we reach adulthood and at different stages of life, when we no longer need them. The actions of genes are not always 'permanent'.

Genes in the distant past may have had different roles. The environments that then existed to trigger gene production of a certain protein may now have disappeared. Alternatively, genes may have been slowly changing as a result of evolutionary pressure over hundreds of thousands of years. In a new environment, some genes will be advantageous or disadvantageous or neutral. A simple example of this is the hedgehog, which for thousands of years has defended itself by rolling into a ball, but has not had time to evolve further – and so now (understandably) has problems defending itself against motor cars.

Another good analogy to show how genes and environment interact in humans is the risk of a nonswimmer drowning in a swimming pool. Let's say that a person's height is 'genetic' and the depth of water is 'environment'. For common depths of water, the taller you are the more chance you have of standing on the bottom. Therefore, at shallow to moderate depths, genes for increased height increase your chance of survival. If the depth of water is greater than 2 metres or less than 10 centimetres, your genes become irrelevant. This is a useful model to consider when thinking about the role of environment for most of the traits and diseases discussed in this book.

In many ways, a genetic predisposition is no different from an environmental predisposition. For example, if you have a poor French teacher at school, your abilities to speak and read French will obviously be less good than those of a student of the same ability with an excellent teacher. However, a change of teacher (an 'environmental change') could dramatically improve your French skills. Similarly, if you inherit a genetic predisposition to find difficulty in learning French, you will have an impaired language ability but improvements in the quality or quantity of tuition can improve your skills markedly. Conversely, if you inherit the same poor linguistic genes, but are never exposed to French, you would never be diagnosed as having difficulty learning languages. Many critics worry about genetic determinism, but seem paradoxically happy with the concept of environmental determinism.

Our genes and our environment are mutually dependent to influence our bodies and behaviours. Without normal stimulation, an eye covered up from birth for several weeks will never be able to work properly. The closest natural examples of environmental deprivation are rare and sketchy reports of human children brought up without human contact, either in the wild by animals (feral children), or totally alone (Kasper Hauser). These children are unable to speak and communicate properly and find walking on their legs difficult. They never catch up with other humans exposed to a normal environment.

So, whatever our genes, our basic surroundings are obviously still crucial to our normal development. Thus, when we say a particular trait is genetic, we are grossly simplifying a statement that the differences between people for this trait are partly influenced by their combination of genes within a particular environment.

2

Genes, Worry and the Early Years

All parents are genetically programmed to worry about their children in order to protect the continuation of their genes. Here, we will look at different scenarios where understanding how and why things happen, and the balance of nature and nurture, can alleviate some of the angst when dealing with young children.

What will my baby look like?

F was overdue, and her anxious husband H rushed to the local hospital as her first contractions began. A number of questions raced through their minds about their first baby. Her previous pregnancy had miscarried last year at twelve weeks. Will the baby be normal? Will F survive the pain of a long labour? Will they need forceps or will her husband faint?

A week earlier their thoughts had been more mundane: Will it be a boy or a girl? How big? Will it have its mother's red hair and pale skin or the father's swarthy Mediterranean looks? Eight hours later, to everyone's relief, baby T was born: a seven-pound healthy boy with a temporarily squashed nose. He had olive skin and dark eyes and hair. Mum immediately said he looked just like his dad. He also had some odd bluish markings around his bottom that looked like bruising. Why did T turn out as he did and which of his parents' features will he exhibit as he gets older?

Like all successful births, T was a miracle baby. The average sex act has only a 2 per cent chance of producing a baby. One of his dad's sperm had met Mum's egg at just the right moment for fertilisation, within a few days of release, and by chance that particular sperm contained not the large X chromosome but the tiny Y chromosome containing the male sex gene SRY. Without this SRY gene, the egg would develop as a female, but the gene triggers other genes and hormones to convert the female into a male. In the fight for the egg, millions of other sperm with different combinations of Dad's genes failed to win the race on the long road up the cervix and Fallopian tubes – and died wastefully in the process. The correct numbers of chromosomes had come together and, despite all the shuffling of the parents' genetic material and the odd new genetic change (mutation) occurring by chance, no fatal genetic flaws were present. These would have resulted in an early miscarriage within a few weeks, an occurrence perhaps affecting 50 per cent of conceptions and usually undetected. Later clinical miscarriages in 15 per cent of conceptions occur as a result of more minor genetic problems. By sixteen weeks, most genetic problems have been weeded out naturally.

A battle of the genes then begins. The foetus (carrying 50 per cent of genes from the father) starts to fight against the mother's body. The foetus demands more and more nutrition at the mother's expense. There is a constant struggle between the selfish genetically programmed needs of the foetus and the mother trying to balance her survival against that of her child – for example, suppressing her immune system so as not to reject the child, and, in so doing, exposing herself to infection. Meanwhile, in the foetal brain there is constant activity. Every minute, 500,000 new neurons are produced that must migrate to the correct location and form connections with their neighbours at the rate of 2 million per second. The brain's trump card, however, is its ability to respond to the endless barrage of information from the outside world by resculpting itself in a lifelong task of pruning old connections and forming new ones, showing remarkable plasticity in the process.

The birthing process is the final hurdle. Control of birth weight is crucial for survival. Historically, if babies were too big or too small they wouldn't survive, and their genes wouldn't be passed on. Nowadays, the increasing use of routine surgical Caesarean births may mean that babies can increase in size in successive generations and live to pass their genes on – such that the average female may not in the future have a pelvis large enough to deliver naturally. This has already happened with certain cattle, bred for years to have such large rumps (for steak) that they now always need a surgical delivery.

Immediately after birth it's hard to discern more than the most obvious of features: all babies' eyes are initially dark blue, some babies are born hairy and others are bald (which, fortunately, are not related to the adult traits). Most people, even paediatricians, can't tell the sex of newborns wrapped in blankets. It has been shown statistically that baby girls are actually hairier on average than boys. Physical features are generally inherited equally from each parent by chance, but surprisingly we still know little of exactly how many features – such as noses, lips, eyebrows or facial dimensions, skin or hair colours – are passed on and which type predominates. African skin-type genes tend to dominate white skin even in dilute amounts after many generations, whereas Australian Native Aboriginal skin type tends to be lost after one generation of mixing. Most of our characteristics are a subtle blend of many of our parents' and grandparents' genes rather than an all-or-nothing scenario. Certain facial characteristics such as prominent noses or chins do tend to run in families for many generations without being diluted, as you might expect. The best example of this is the Hapsburg lip, a protruding lower jaw that was first seen in a German nobleman in the tenth century and carried on through the European monarchy for nearly nine hundred years.

Sometimes small skin markings reveal clues of our family origins. The small bruise-like stain on the lower backs and bottoms of babies is called a 'Mongolian blue spot'. It is very rare in northern Europeans, but quite common in Asians. Some parents have been charged with baby battering because of these marks – which are merely evidence of Asian ancestry (perhaps many generations ago when Attila the Hun and his hordes reached as far west as Paris)'.

Despite all the recent progress in genetics we cannot predict with any certainty what parental characteristics our babies will display. This ensures that the genetic lottery of life remains, in most cases a pleasant surprise. Nature demonstrates this dramatically when some nonidentical twins are born at the same time but are very different in appearance. Sometimes, with mixed-race parents, one twin baby can even have black skin and the other white.

Studies of modern births have shown that, soon after delivery, 75 per cent of mothers declare that the baby looks like the father. This may be instinctive behaviour to protect the child in case the husband suspects he may not be the father. Infanticide may have been an unpleasant part of our ancestral past and is seen in hunter-gather societies, where mothers sometimes decide which babies to keep and reject others on the basis that they have poor chances of survival – either from ill health or from the father. Alternatively, babies may be programmed to look more like the father in the early stages to protect themselves against the wrath of an angry father who thinks he may have been cuckolded. Why babies are inefficiently born so fat and chubby and hard to deliver is also a bit of a mystery, but may relate to the same problem – convincing the mother that they are healthy and worth looking after. Once they've found the mother's breast, babies are much safer and much less likely to be abandoned.

Will our next child be a boy or a girl?

After three boys in a row J and G were still keen on having another child, but were really desperate for a girl. J himself had three brothers and no sisters and his wife's only sibling was a brother. Were they destined to have only boys? Could they produce a girl by careful timing of intercourse, or was there a special predictive test or procedure they could undergo to increase their chances?

In most mammals the sex ratio is close to 50:50. Nowadays among humans slightly more boys (2–5 per cent) than girls are born. In the UK the ratio is currently 51.4 per cent boys. Until very recently boys had much lower survival rates as babies. Producing males for most species has always been a riskier business for mothers than producing female offspring. If the mother isn't in a state of good nutrition and health, boys will be born more susceptible to disease and early death than girls. Recent studies have shown that mothers of male babies eat 10 per cent more than those of female babies – this is because the male baby signals the mother via testosterone that he needs more energy. Because, later in life, they also need to be in good health to succeed and reproduce, males are therefore a higher-risk strategy. Females are a safer bet, as they are more likely to find a partner, even if their health or circumstances are impaired. In nearly all societies and cultures there are more men who fail to reproduce than females. In the children of starved mothers in Holland at the end of World War Two, death rates were higher in male babies and infants. Later on as adults, those who survived were also less likely to marry than their sisters.

In many developing countries there is considerable pressure to produce male children at the expense of 'costly' females. This is partly because of the (albeit illegal) dowry system and partly due to cultural values. Infanticide of females is quite frequent in many parts of the world, such as India, where the male-to-female sex ratio of recorded births in

1991 was 108:100 and a study of eight thousand terminations found all but three were of female foetuses. In China, where couples are allowed only limited families, the overall ratio is 117:100 and it was estimated that in the 1980s around fifty thousand girls were killed annually. Modern technology, such as ultrasound scanning, is helping detect gender at an earlier stage. Even in poor countries, paying for this service and the resulting possible termination is seen by many as good value for money. Richer prospective parents can rely on even more hi-tech methods such as in vitro fertilisation to ensure a child is of the desired sex.

In developed countries more subtle methods of selection are used, such as the 'male sperm swim faster' theory that urged couples to copulate on the day of ovulation for boys and a day or two after for girls. This works miraculously half of the time! Another traditional method (with the same success rate!) is for men to lie on their right side during sex in order to have a boy. Based more on science than astrology, clinics in the US now offer services at around $2,000 a time (also available on the Internet) to sort the father's sperm. This method separates male sperm carrying the Y chromosome to produce a boy, from those carrying the X chromosome to produce a girl with reasonable accuracy (about 85 per cent), so that only single-sex sperm are inseminated. In most developed countries there are no major gender biases. However, in the US, for couples wanting sperm sorting, the current trend in gender preference is the opposite to that in poorer countries: girls are preferred to 'high-risk' boys, perhaps because they incur fewer medical bills early on, or are a better investment for future social support.

Is it wise to tamper with nature? Studies of animal populations have shown that, within a few generations, the balance would naturally revert to the 50:50 ratio that evolution has decided suits us best. After World War One in Europe, rates of male births increased temporarily, then decreased, to rise again after World War Two. In China, a lack of females is having a harmful effect on society, with increasing numbers of abducted females and violence by gangs of bachelor males who have little hope of marriage.

We are aware of a few factors that slightly affect gender ratios: more boys will be produced to nonsmokers, mixed-race couples and young couples; chemical contamination can also reduce the numbers of boys. In the US and other industrialised countries, the male–female ratio has decreased slightly but significantly over the last twenty years to around 1.047:1. This slight difference results in 37,000 fewer baby boys now being born annually in the US than in the 1980s. Studies have shown that more boys are born in southern Europe than the north and fewer in Mexico than Canada. These differences remain unexplained. Whether some families carry genes that predominantly influence gender is not yet known. The many stories of bizarre families, with, for example, 76 daughters in three generations and no sons, may still be due to chance. These have to be viewed within the context of the billions of families worldwide for most of whom the sex of a baby is a pleasant 50:50 lottery.

He has the milkman's eyes

After about six months of age baby H started to reveal his genuine traits. What hair he had was a light brown. His eyes, which at birth were dark blue like those of other babies, were now definitely bright blue and quite striking. Mrs B was very happy until a 'helpful' friend of hers remarked over coffee that she thought the baby's eye colour was odd – because Mrs B and her husband both had brown eyes. She had heard that brown eyes were always dominant, and if both parents have brown eyes, there was no chance that any other colour would appear, especially blue. The unclear identity of the father was subtly insinuated. Although H's mum knew she had been faithful, the thought that people might gossip about it upset her.

Most people assume we know exactly how eye colour is inherited. Actually, we know surprisingly little, and the subject is much more complicated than we imagined. All our ancestors' eyes were likely to have been dark brown, perhaps providing protection against bright sunlight. As man spread out into darker and colder environments, genetic mutations for eye colour occurred. The coloured pigment (called melanin) in the eye controls the colour: those eyes with a lot of the pigments are brown and those with very little are blue – the green, grey and hazel being intermediate shades.

There are two principal genes controlling eye colour, each with two possible forms: the *bey gene* controlling blue (b) and brown (B) on chromosome 15 and the *gey* gene controlling green (G) and blue (g) on chromosome 19. Your eye colour is dependent on which of the four forms you have. You inherit two from each parent – so four in total.

There is a colour hierarchy between the different forms of the two genes, resulting in some simple rules whereby some colour genes are stronger than others:

- B is always dominant over b, G and g: if you have just one brown form and all others are blue or green, you'll have brown eyes;
- G is dominant over b and g: if you have one green form and the rest are blue, you'll have green eyes;
- b and g are never dominant: you'll have blue eyes only if all your gene forms are for blue eyes.

For the baby in the scenario to have blue eyes, both brown-eyed parents must have each had one brown copy (B) and three blue copies (bgg) and produced a baby who had all four blue copies (bbgg). Equally, one or both parents with brown eyes could have the BbGg combination. This simple explanation based on two genes works for most families, but exceptions still occur, and it is known that other genes exist, such as the BEY1 gene controlling the central brown colour on chromosome 15, which can modify the brown colours and

BG	bg	BbGg	(Brown)

Bg	bg	Bbgg	(Brown)

bG	bg	bbGg	(Green)

bg	bg	bbgg	(Blue)

The 'commonly observed' patterns of inheritance are: two blue-eyed parents = blue-eyed children; blue + green = all green or 50 per cent green, 50 per cent blue; blue + brown = all brown or 50 per cent brown, 50 per cent blue; green + brown = all brown or 50 per cent brown, 50 per cent green or blue.

shades of the eyes and cause more confusion. We still can't account for eyes that are grey or hazel. Nor does it explain how two blue-eyed parents can rarely have a brown-eyed child or why eye colour can change over time. It's likely that many more genes than this are involved

So, if you are a worried father, read up on your genetics before attacking the milkman!

What's in a name?

Just after the exciting event of the birth of their son, the proud parents M and B tried to agree on his name. For months they had been arguing – and as yet had not agreed. They were

down to a shortlist of ten or so. The problem was that the family had a rather unusual surname that made it harder to choose a name they both liked. As they were about to leave the hospital, a nurse came up to them announcing she had the same surname and wondered, since it was rare, whether they were related to each other. They swapped what information they knew, but neither knew much about their grandparents' origins, because they had died – and the mystery remained unsolved.

Have you wondered whether you are related genetically to other people of the same surname? One genealogy enthusiast with plenty of spare time calculated that within the Windsor family Prince Charles had at least 262,142 ancestors. An Oxford geneticist called Sykes performed a more down-to-earth study in England. He traced two hundred apparently unrelated males with the surname Sykes. This name is a quite common and ancient name in northern England, having been around for about seven hundred years. It was probably the old name for ditch. He examined their DNA and in particular a set of markers on the single Y (male) chromosome, which is always passed from father to son, finding to his surprise that, of the sixty men tested, the majority had similar segments of Y chromosome and were related via a common male ancestor.

In most areas of the world the number of different names in a town or country gives a clue to the variety of different genes and therefore the amount of mixing of populations in that area. For example, the White Afrikaners have only about twenty names, suggesting that they are descendents of only about twenty families who migrated from Holland in the seventeenth century. In England, geneticists successfully matched the DNA of a 9,000-year-old skeleton of the so-called Cheddar man with the DNA of a local schoolteacher living near the cave where the bones were found. A group of second-generation West Indians living in the UK were recently traced back to ancestors in villages in Equatorial Africa two centuries and eleven generations before the slave trade took them to Jamaica. The popularity of seeking out one's roots is best seen in the obsession of US presidents to find links with the old country, nearly all managing to find key Welsh or Irish relatives. Ireland has even taken the credit for the boxer Muhammad Ali's prowess, attributing it to some mysterious Irish genes.

Net surfing for genealogy is one of the fastest-growing areas of the web (second only to sex). Several DNA testing services (Relative Genetics, Oxford Ancestors, Family Tree DNA) have now been set up in the UK and USA to test for common ancestors in potentially related individuals for a cost of £100 ($150) a test. One was set up by a Jewish entrepreneur, who found there was a niche for people seeking relatives, especially in certain groups when surnames have been used only in the last two centuries or when migrations and wars have altered the spelling. He found, for example, that, of 570 males with the name Katzman in North America, most were related and he could construct six distinct family trees. Sam Zaidins from Palm Springs found a Russian émigré in New Jersey called Mickael Zaydens. They had the same Y chromosomes and turned out to have descended from second cousins. Names can be fickle as clues to our history, for some have only been around for a short while, or changed suddenly due to emigration (such as the Americanisation of refugees' names at Ellis Island, New York). Others became slowly mutated as the spelling changed. Some names became dead ends as they ran out of sons or were genetically diluted by infidelity.

It's obvious we are all related to *some* degree. It's now believed that most current Europeans are descendents of one of perhaps only seven original families (or clans) of modern man who crossed from Asia into Europe for the first time around fifty thousand years ago. Taking samples from the whole world, it appears we are all derived from around 33 clan families of which a large proportion,13, are clearly from Africa, our home for most of man's history. Of course, these 33 clans had to be descended from someone. The evidence points to an African female who has been named Mitochondrial Eve, who lived 150,000 years ago and whose female descendents kept the line going. In other words, there's no such thing as a total stranger – particularly if you share a similar name. We can safely say with reasonable certainty that it's likely we are all descended from people who had lots of children – such as Nefertiti or Confucius.

Cot death

Mrs C was in the hospital accident and emergency depart-ment as the news was given to her of the death of her eighteen-week-old son. She was too shocked to notice the presence of the policewoman behind her and barely heard or understood as she was told she was under arrest for murder. This was Mrs C's third child and the previous two had also died in suspicious circumstances eight and nine years previously. They both had episodes nine days prior to death when they were very ill or stopped breathing. They had been called 'cot deaths'. She had been given a breathing alarm, but had apparently not heard it when her third baby had died. At her trial the main prosecution witness, a highly qualified expert on cot death, told the jury that one death was unlucky, two highly suspicious and three murder. The odds of this happening by chance were just not credible – he estimated that the odds of one were 1,850 to 1, of two 73 million to 1, and of three 130 billion to 1. She denied the crime but was sentenced to a mandatory double life sentence.

Cot death (or sudden infant death syndrome) remains a medical mystery. For unknown reasons some children just stop breathing and the term is used when no obvious cause is found. It has been estimated that around one baby dies like this every day in the UK and about five per day in the US. The risks are partly related to the baby's position in the cot – lying on the back may reduce the risk, and overheating and exposure to cigarette smoke are risk factors. There are unexplained hundredfold differences between countries but genetic factors are difficult to study, as the problem is so rare. In most cases doctors can't tell the exact cause of death and the features are usually indistinguishable from smothering and suffocation. To complicate matters, infanticide (which is mostly carried out by the mother) does sometimes occur in modern society, where it has become a taboo subject.

Exactly how often infanticide is the cause of death is guesswork. Some studies, mainly from the US, have shown rates of 3–6 per cent are definitely due to murder – but some experts believe the true figure may be as high as 40 per cent. Women in primitive societies are reported to commit infanticide often when the conditions are not ideal for the survival of their child – such as when they have changed partner or food is scarce, or where the baby is born under-developed or with a deformity. Chimp females also commonly practise the same tactics, especially when there are high risks that a new male will kill the baby anyway. It is possible that some modern women still carry the instincts and ability to commit this crime. These instincts may be switched on if the mother feels there is a major threat to her own survival or to future children – horrific though it may be to contemplate.

In Mrs C's case, other experts have begun to question the expert evidence that cot deaths cannot occur more than once without infanticide being likely. Mrs C's legal team consulted other experts who were geneticists and mathematicians not used in the trial. They disputed the original claims that the odds were millions or billions to one as ridiculous – as the causes (even if unknown) were likely to be related. One expert from Canada estimated that there was a 25 per cent risk that subsequent children would have the same disorder and die suddenly in infancy. Another geneticist traced Mrs C's extended family of second cousins (sharing one-thirty-second of their genes). They found two other families with two or more unexplained cot deaths in each family – suggesting the possibility of a rare genetic cause. The search is ongoing for more relatives.

Some recent studies of the parents and siblings of cot-death cases have found suggestions that their genes may be different from those of normal families. The genetic differences are in genes of key areas affecting response to infection or affecting chemicals in the brain (serotonin) that cause alterations in temperature or breathing. Others have found mild defects in the mitochondrial DNA, carried only by females, which are very difficult to detect. In another recent high-profile case, a Mrs P was acquitted by the jury of the murder of her three infants – after hearing of unexplained deaths of other children of relatives in India. Based on new medical evidence, Mrs C has started an appeal. Meanwhile, like others, she is in prison – either as a child murderer or an unfortunate carrier of a rare genetic disease.

Parenting instincts

A was seventeen when she unexpectedly missed her period and found out she was pregnant. Her eighteen-year-old boyfriend of six months was shocked when she told him. They discussed what to do and agreed she should have a termination. A week later she changed her mind and wanted to keep the baby. Her parents and family thought this unwise – and doubted whether she would cope mentally and physically. A and her boyfriend had an acrimonious discussion, and a few days later he left both her and the city. Contrary to expectations, A brought her child up on her own and was generally a good and caring mother. Her son never saw his father.

Women, like other mammals, have maternal-instinct genes that make them care for their children. Otherwise, as human babies are so helpless, our species wouldn't have survived. Genes control the desire to reproduce and perpetuate the genetic line, protecting the children. Like all genetic traits, they vary, and some mothers are more successful than others. Evidence from the animal kingdom shows that, in the majority of cases, the female of the species takes more responsibility for raising children. This is overwhelmingly true in mammals with very few exceptions. It boils down to relative investment. Females invest vastly more resources in producing a child, from the minimum time and energy involved in producing a large fertile egg, to growing the embryo, and then to breast-feeding it for at least three more years.

The evidence that paternal instincts in fathers are genetically influenced is seen from studying animals where a very wide range of fatherly behaviours is seen – from instant abandonment to lifelong devotion. Wolves have strong parental instincts but, after eleven thousand years of intensive breeding, produced domestic dogs with a total lack of paternal-instinct genes. Presumably, those that retained this instinct objected to their families being separated. Fish apparently make good fathers, but this is because the female lays her eggs first and has time to run away before the male can cover them with his more fragile sperm, thus leaving the father holding the baby. Given the choice between investing time and the certainty of letting his progeny (genes) die, the male fish chooses to stay. He also gets rewarded by often attracting other females who are impressed by the eggs and his commitment to them. Perhaps this is why single-parent human males can also be attractive to other women.

Our close relatives the apes are useless fathers – they lose interest straight after the birth and often never see their children. The main reason apes and many humans are not better fathers and invest more is that they can never be 100 per cent sure they are the father. Most female apes are notoriously unfaithful – having sex with up to twenty different males in a day. Being an unknowing devoted father to some other male's child is a disaster avoided by male apes, who show no interest in their offspring. One of the few examples of great mammalian dads are the titi monkeys. They tend to be a close faithful couple, who sit on branches with entwined tails and share the childcare duties. Their closeness ensures her fidelity and therefore his commitment.

Why did humans develop paternal instincts, which most apes lack? One reason may be that, as humans started walking upright, females had to carry their young. If they were left alone they often had trouble collecting as much food – and risked going hungry. They therefore needed men to hang around and provide support – and so they started choosing men on this basis.

The parent–child bond that develops – the so-called 'parental warmth' – depends equally on the genes inherited by the parent as well as the child. This was confirmed by a detailed psychological study of 1,400 parents of twins in the US that showed the importance of genetic factors on both the reactions and perceptions of parents and children. Warmth involves levels of hormones produced by the body. One of these hormones is oxytocin, which is responsible for the feelings of wellbeing and familiarity. The gene influencing response to this hormone (oxytocin receptor gene) may be important in determining different responses and bonding. Different genes and chemical clashes between individuals may explain why some parents bond better with one child than the other, despite identical environments, upbringing and the best intentions. The other aspects of child–parent bonding are protectiveness and authoritarianism. Neither of these was found to be particularly genetic and both appear to be dictated more by our environment – as is clear when we look at our Victorian ancestors.

Human babies have evolved and developed ways of making sure their parents look after them. These tactics include very early eye contact – despite poor vision at birth – as well as copying parents' facial signals and early attempts at smiling. Babies also produce subtle chemicals called pheromones, which can subconsciously signal other humans via the air. These are particularly directed at fathers in the first few months, who in experiments detect them more easily than mothers. They are perceived as pleasant and calming influences. This may be a mechanism to increase the chance that Dad will stay relaxed and calm and be around for the first few months when the baby is most vulnerable to male traits such as abandonment or violence.

In Europe, where birth rates have steadily fallen since the 1960s, the average number of children in the UK is around 1.7 per woman and closer to 1.0 in parts of Italy and Spain. Around one in four couples do not have children and most of this is now a result of parental choice rather than infertility. The accusation by the majority of parents that childless couples are in some way 'selfish' is curiously hypocritical, because the evolutionary drives of parents to *continue* their own genetic lines are programmed to be selfish. Although contraception is relatively new, our ancestors managed partly to control their child-rearing and spread out their pregnancies to every four years; the same is true of recently discovered primitive tribes. Man's control of such a natural event is a good example of free will winning over millennia of inbuilt instinct.

However as the birth rate continues to decline, there will be fewer members of succeeding generations to look after their parents. This problem is particularly acute in Europe and Japan but may eventually affect the rest of the globe. It is now predicted that the world's population will have peaked by 2050 and start to decline – although, with 6 billion of us at present, it's going to get even more crowded first.

Why incest is not best

C met her handsome cousin at a wedding and felt a little buzz of excitement. They had always got on well. They met on a nearly annual basis for the last ten years – but this year she felt something different. They danced together for the next few hours, and under the influence of alcohol, became increasingly intimate and aroused – and later had sex at their hotel. They both felt a bit guilty about their relationship.

Even though they were both over 21 they knew that cousins were not really supposed to be lovers. To C's surprise, a month later she found out she was pregnant and they decided to marry. They were delighted when a baby boy was born – but were distressed when six months later he was diagnosed with cystic fibrosis. They were informed this was a special genetic disease, caused by a mutation in a single gene, and that their baby had an increased risk of the disease because they were cousins.

Incest has been taboo for as long as human history, the only exceptions being humans who were considered to be gods, such as the Egyptian pharaohs and the Inca royalty, who often married their sisters and half-sisters. Most animals in the wild also have systems to prevent incest. Incest with a first-degree relative (siblings, mother, son etc.) is a definite taboo in virtually all human cultures, as the sharing of genes is already 50 per cent. There are environmental and genetic reasons for this taboo. On the environmental side, even if you wanted to mate with your parents or siblings, having to bring up the resulting kids within the existing family circle would not be a sound social or economic plan, as the number of spare hands would be diminished.

The biological reason for the taboo is that we all carry a few defective genes on our chromosomes, the effects of which are usually nullified by the working copies of the same gene on the other paired chromosome. Any particular risky gene may be carried by, say, one in a hundred of us and normally the chance of a random couple sharing a particular defective gene is small – around 1 in 10,000. If a couple are related, they are more likely to share the same bad genes at the same position on the chromosome, and the risks of having a baby with a disease or dangerous trait are much higher. For brothers and sisters there is a one-in-four chance they will be carrying the same potentially lethal genes – resulting in a very high death rate of any offspring. In many cases nature deals with this by causing a miscarriage, but if the genetic defect is more minor a child may survive who carries both copies of the bad gene and develops the disease.

In human society the marriage and mating of kin is tolerated at the level of cousins, although it is still illegal in a few countries and states. This level of genetic sharing means they share one grandparent and so 12.5 per cent of their genes. In the past, with small communities and less choice, cousin marriages were very common and often a reasonable choice in the circumstances. It was often a way of keeping land within the family and was, for example, commoner in the southern US than the industrial North. Today rates in most Western countries are fewer than one in two hundred marriages, but are still up to one in three marriages in those living in remote mountain villages such as in Italy, the Balkans and Switzerland, as well as in the Middle East and in arranged marriages in immigrants. Overall about 1 billion humans live in societies where cousin marriages are common (more than 20 per cent). The risks to their children are slight but measurable, the chances of any genetic defect in the offspring of cousins being slightly higher than the 5 per cent in the general population.

Cystic fibrosis is the most common single gene disease in Western countries, causing lung and breathing problems. It is called a *recessive genetic disorder* – as you need two copies of the gene CFTR to cause problems. It is relatively rare – affecting one in 2,500 people – yet the genes are carried (one copy only) by about one in ten outwardly healthy Europeans. It is usually caused when two carriers marry, and the chances of their both being carriers increases the more a couple are related and when they are from similar genetic stock. It can be tested early in life by biochemical tests on a baby's sweat, and genetic testing before birth is possible, though quite difficult, because the CFTR gene has many different abnormal forms.

Although rare, there are thousands of recessive genetic disorders and we all carry one copy of some of these potentially dangerous genes. Many of these genes had some useful function in our history at some time, such as protection against a deadly plague or parasite – which explains why evolution didn't get rid of them. Today, in our new environment, many are now redundant. Others, like sickle-cell disease, still protect carriers against malaria. Carriers of one copy of the cystic-fibrosis gene (who are otherwise healthy) are protected against the worst effects of stomach infections and diarrhoeas caused by bacteria, a common cause of death in harsh climates.

We only have to look at a few pedigree dog breeds or the Russian royal family to see the dangers of forced inbreeding that commonly leads to blindness, deafness, early arthritis and haemophilia. To prevent this happening accidentally in humans, we have evolved both cultural and genetic checks. Our bodies evolved ways to detect related family members by various methods, including smell, and we can both send and receive anti-sex signals. Experiments on animals have shown that females inseminated simultaneously by a brother and a stranger nearly always reject the relative's sperm. Studies of 2,769 children brought up together in Israeli communes have shown that the children we play closely with from an early age (before the age of six seems critical) are virtually never the ones we end up marrying. We grow up treating these playmates as if they were kin. So, if you want to marry your childhood sweetheart, make sure you wait until the tender age of six to work your charms!

'Designer babies'

M and J W had a four-year-old son, C, with a rare blood disorder – anaemia – which is usually fatal and requires long, painful treatments. He desperately needed a bone-marrow transplant (containing healthy stem cells) from a relative. But none of the family members could provide a good enough match. The family asked whether they could have IVF treatment to produce another child that would at the embryonic stage be selected to provide a perfect genetic

*match for the bone marrow of their other son. The UK
authorities refused. J and M went to a clinic in Chicago for
the IVF, where only one fertilised egg (of many) was
selected, which had a 98 per cent match with C's key genes
in his bone marrow. Nine months later in England, a little
boy, J, became in 2003 the country's first 'designer baby'.
Time will tell whether his creation will save his brother – or
whether he too will also succumb to the rare anaemia that
runs in the family.*

In most countries the use of stem-cell research and genetic
selection to directly save a life is accepted in principle – but
often banned in practice because of a lack of clear guidelines
about the selection process and problems in regulation.
However the real-life scenario above is likely to become more
commonplace over the next few years. Implanting genes into
humans and changing for ever the germ line (i.e. DNA of
subsequent generations) is another stage further. It is now
close to reality, although remains banned in most countries.
Recently, a monkey named ANDi (short for 'inserted DNA'
backwards) was produced by genetic engineering in a
landmark experiment in Oregon. A special jellyfish gene that
normally codes for a green fluorescent protein was inserted
into his DNA at the fertilised-egg stage. This first genetic
implant in a primate had no obvious harmful effects – but may
have made him appear brighter. There are many other
examples of gene implants working in less closely related
animals, and artificial chromosomes are now being constructed. You can now buy genetically modified pet fish that
glow in the dark in their tanks.

Gene therapy has been around since the late 1980s, but
with frustratingly little success in humans, apart from the case
of one very rare disease called X-SCID, where children lack
the genes that produce certain white cells that protect against
infection. Gene therapy has worked in several cases with this
disease, but has caused side effects of leukaemia and allergic
reactions that may also be fatal.

Most of the problems with gene implantation are due to the way genes are introduced into the body using viruses. It's hard to control where these viruses travel and where they replicate. Advances in technology are also occurring, such as artificial wombs, so that the early growth of the egg and foetus can take place under laboratory conditions. While we can now in theory select embryos carrying certain genes, we cannot yet modify them. It is not clear how long it will take, but scientists are excited by the future prospect of inserting genes, maybe without viruses, to prevent breast cancer, epilepsy, schizophrenia or infections such as HIV or malaria. This would have obvious benefits and increase longevity of humans, but what of the dangers? The likely costs will be socially divisive. Should you be able to do this to avoid criminal or antisocial behaviour, or enhance intelligence or athletic skills?

Cloning is another technique that is now with us to stay. The first animal clones were produced in the 1980s, initially with frogs and then with small rodents, progressing to the first large mammal, Dolly the sheep, cloned in Scotland in 1997 from her mother, a domestic cat, 'CC', in 2002 and more recently a female horse gave birth to her cloned twin 'Prometea'. Cloning in humans is now feasible, but has been banned in most countries for reproductive purposes after considerable controversy, and debate is ongoing. Cults such as the Raelians, who believe we are all cloned from an extraterrestrial, have, however, claimed to have successfully carried out human cloning – without producing the evidence. Needless to say, it is a major area of controversy and extreme views, partly related to wacky Hollywood-style scenarios, such as cloning famous historical figures from their hair.

One argument to support cloning is that it can produce novel treatments using the body's own building-block cells, called *stem cells*. Cells from the target patient or donor are taken and grown, and the genetic material (DNA) is removed. This is then added to a hollowed-out female egg, which contains none of the female's genes. The two are fused together by a quick burst of electricity, and the new egg with the new genes starts to grow. If left for nine months, it could in principle turn into a baby clone of the father, separated by two to three decades, with no genes from the mother.

The medical benefits of cloning lie in being able to take the grown-up stem cells and alter them to produce chemicals that the patient lacks, such as insulin in diabetics. Missing tissue such as the kidney can be grown and the functioning cells implanted back in the patient, where they can function normally and replace the original abnormalities. The reason the cells need to be cloned is that the body would then attack any new cells it thought were foreign. No one knows whether cloning in humans will be a safe, healthy procedure. It took more than two hundred cloning failures to produce Dolly, and after three years she developed arthritis at an early age, and died aged six of lung disease, whereas most sheep live for eleven years. This suggests that she was not as healthy as believed – perhaps she aged faster or had other hidden problems related to being a clone. Similar health or ageing problems have been found in other cloned species, particularly in apes, suggesting our DNA may be more sensitive than we thought.

Nevertheless, stem-cell research is continuing in many countries with encouraging signs of early success in a few areas. For example, cells that can grow new kidneys when injected into an animal have now been successfully produced. To many people, the idea of having a clone is a frightening idea, and many believe cloning should never be allowed to happen. Yet today, secretly walking among us in the world, are around 20 million clones. They're called identical twins and they don't seem to mind.

3

The Early Years

Most parents find the early childhood years – from the time babies learn to walk to the time they go to school and become more independent – to be the most enjoyable. During this period, the child's brain is growing fast and adapting to its environment. Psychologists believe this is the crucial time when future personalities are shaped and moulded with good and bad habits being imprinted.

Sugar and spice

C's parents were quite surprised when their two-year-old daughter grabbed a doll from another toddler and started playing with it in a very maternal way. Being up-to-date, well-educated parents, they had always believed in not stereotyping their children with gender-specific toys. Despite their best efforts, their four-year-old boy was much more interested in trucks and toy weapons. They felt sure they hadn't consciously influenced their children's preferences – but maybe their children's choices were the effect of TV programmes, commercials or other children.

Since the 1960s, psychologists have been telling parents that the way children are brought up determines their behaviour. Females copy their mothers and males their fathers. We were led to believe that, given a neutral environment, boys wouldn't always pick violent weapons and girls wouldn't play with toy babies. In other words, nurture ruled. We now know this is wishful thinking. A few rare unfortunate examples of normal children who changed gender have altered our perceptions.

A pair of identical twin boys were born in 1963 in the US. Both were quite normal and healthy when hospitalised aged seven months for a routine circumcision operation – but John's operation went horribly wrong and an electric cautery machine burned off his penis. Surgeons advised the parents that they could make John a vagina much better than a new penis. After guidance from counsellors who believed gender identity is learned, and several lengthy operations, John became 'Joan' and was brought up as a girl. John's parents made him wear dresses and play with dolls and he was treated in all respects as female by friends and family alike. Joan was miserable and friendless, always wanting to pee standing up, and wanted to wear trousers and climb trees. When, aged fourteen, 'Joan' became suicidal, her parents finally told her the truth. She took male hormones, had a mastectomy, had a new penis made and worked out with weights, and the reborn 'John', aged 25 married a normal female. This and other stories show the power of genetic instinct (nature) over nurture in gender.

Some psychologists still believe that gender differences in young kids are mainly due to parental influence. However, summarising 172 studies of parental behaviours across all cultures and countries, researchers could find no real differences in the way parents treat young boys or girls, other than in the purchase of toys. By the age of two, most children have clearly shown some gender-specific preferences, with research showing that girls like to choose relationship-type games and boys prefer action-type games. That genes control male/female differences even earlier in life is becoming clearer. Studies of foetuses still in the womb have shown that, even after a few months, males have longer arm bones than females – likely to be useful for throwing and hunting. Within 24 hours of birth, boys look at objects more than girls, who prefer faces. By the age of twelve months, boys make less eye contact than girls and choose different toys. By the age of two or three (before they've had much practice) 90 per cent of boys can throw a ball further than girls. Differences in these traits have been partly linked to the extra testosterone levels in the womb that boys are exposed to, which, even in infants, can produce regular erections. We discuss later the role of exposure to hormones and genes and upbringing in the more complicated situations such as homosexuality. Transsexuals and transvestites present an even rarer group of human behaviours – and the causes are still unclear.

While boys have the advantage in gross physical abilities, girls tend to develop verbal and communication skills much faster than boys – an advantage that persists into adulthood. Strangely, it appears that a gene on the X chromosome passed on from the father determines that the girl will have the extra emotional, social and communication skills and female behaviour that her brother would lack. Turner Syndrome girls have a mild genetic defect caused by the loss of one of their two X chromosomes, making them act more like disruptive and inattentive boys. The most badly behaved girls have retained the single chromosome from their mother and lost the one from the father. This normal protective calming effect of the X chromosome (even if it is the father's version) is one reason why extreme behavioural problems are so much rarer in females. Perhaps evolution couldn't allow females, with their heavy reproductive responsibilities, the freedom to behave as erratically as males.

Gameboy eyes?

Since the age of three, M's children had been avid video and TV fans. At age six, C, the older boy, progressed to small-screen Gameboy games. He was never very interested in sports but became rather obsessed with his games, and would play for hours on end in his room or in the car, often in poor light. At age eight, C, although brighter than average in his class, started to perform badly at school. A year later, his school nurse found he was very short-sighted and needed glasses. His dad had needed glasses from the age of fourteen, as had his uncle. Should C's parents have been stricter – or were his eye problems inevitable?

Short-sightedness (myopia) is genetically programmed. Family and twin studies by our research group at the Twin Research Unit at St Thomas' have shown that genes explain 85 per cent of the differences between individuals. This is partly due to changes in the natural lens and partly to the increased growth and changing shape of the eyeball. The hunt for the genes has been narrowed down to a few areas on chromosomes 3 and 11. Long-sightedness and astigmatism are also similarly genetically influenced, but the genes may be different. In the last thirty years short-sightedness has increased rapidly in children of all developed countries, affecting up to one in three people as kids switch from outdoor to indoor pursuits and use their eyes more for close work. In Asian countries it has now reached epidemic proportions: up to three-quarters of teen-agers require glasses. These changes cannot be due to alterations in our genes, which would take many generations, and must therefore be due to a changing environment. It looks likely that the responsible genes were around for many thousands of years and didn't cause much of a problem until man started to read and do close work.

Short-sightedness was unknown up to thirty years ago in traditional Eskimos (such as the Inuit), who had an exclusively outdoor life. Their survival depends on the ability (particularly of the men) to see animals such as polar bears on the horizon. Without books and the need for close working, long-sighted genes must have predominated in our ancestors, and would have enhanced the skills, particularly of men in hunting.

If long-sightedness was crucial to survival, how did short-sightedness genes persist? It may be that these common genes conveyed some advantage. For some time it was believed that children who wore glasses were brainier and were less interested in sport. But we don't know whether this is because short-sighted children are *born* brainier or because children who study more become short-sighted. One recent study tried to resolve this by examining the activities of parents and children in normal and short-sighted families. Findings confirmed the twin studies, showing that genes are the most essential factor, but that close work – such as playing computer games, watching TV and reading for work and pleasure – were also important in susceptible people. Of these, TV is likely to be the biggest culprit. On average, individuals in the industrialised world devote three hours a day to the pursuit – half of their leisure time – and an average of nine years over a lifetime. Seven out of ten teenagers in surveys said they spent too much time watching TV. In the US, the average is nearly 1,500 hours per year of TV compared with only 99 spent reading books.

The study also established that children with glasses are brainier, read more and do less sport, even after accounting for their parents' genes. Perhaps our rare short-sighted ancestors, who must have had major disadvantages in the wild, were tolerated by their companions – allowed to stay in the camps and caves – and, being brainier, made tools instead of hunting. Perhaps these short-sighted males also stayed nearer the camp with greater access to sympathetic females, and – unknown to their more sporty hunting colleagues – kept the genes going in the traditional way!

If short-sightedness is present in only one parent, the risk in childhood of needing glasses is one in five. If both parents are affected, the risk is one in three and, if neither has it, only one in twelve. These risks are further increased in Asians. If you wish to minimise the risk or delay the onset in susceptible kids, try to encourage outdoor pursuits and activities involving looking further than six metres away as much as possible, limit the duration of use of games or study that require intensive close work and check they always work with good lighting. If this fails and they need glasses, tell them how lucky they are to be so brainy.

How tall will he be? Greens or genes?

Little S had always been shorter than his friends and class-mates. He seemed to eat normally, although a bit erratically. He appeared quite healthy. By the age of nine he was at least 1 foot (30 cm) shorter than all his friends and girls in the class. His parents, who were of average height, became worried and took him to the doctor. The doctor asked many questions about his diet but could find nothing wrong. He told them not to worry and to come back in a year. A year went by and he still wasn't growing. His mum wasn't happy and remained concerned. She had read on the Internet how there are now a number of special diets and expensive treatments including growth-hormone injections, that could increase his height.

Height is one of the most well-known and clearly genetic traits, with over 80 per cent heritability, reaching 90 per cent in males. However, the influence of genes has often been masked by important historical changes to our environment – mainly our diet. From studying skeletons, we see that our ancestors of fifty thousand to a hundred thousand years ago were probably of similar height to us today, probably because as roaming hunter-gathers they were healthy with good diets and plenty of vitamins. When farming started ten thousand years ago and we stopped roaming, our diets took a turn for the worse and lacked many essential vitamins and minerals. One consequence of the inferior diet and possibly extra disease was a reduction of average height by 8 inches (20 cm) in men and women. Average height has fluctuated since, with a further reduction during the Industrial Revolution (in the seventeenth and eighteenth centuries) when many people moved into big cities. During the American Civil War, 150 years ago, the average height of soldiers was only 5 feet 4 inches (162.5 cm). Dutch military recruits were 5 feet 6½ inches tall (169 cm) in 1900, but 5 feet 11¼ inches (181 cm) in 1990 and the Dutch are now the tallest in Europe. In the last hundred years, with better health and diets, average height is now increasing by an inch (2.5 cm) a generation – until we reach our full genetic potential again, after ten thousand years of malnourishment. Even when genetically matched groups such as the Koreans, who were separated artificially into North and South fifty years ago, have different diets, they already show about 4 inches (10 cm) in average height difference between them. Similarly, a pair of identical twins developed, between the ages of eight and twelve, a major height difference when one girl started intensive gymnastics – and her growth rate slowed.

The message for little S and his parents is, as long as he's growing and well, don't worry. Many children have late growth spurts and continue growing past other kids who stop in early puberty. Forcing food on them doesn't help, because they eat when they need to, and a balanced varied diet rather than a high-protein one is better. S's height will on average be related to both his parents with a 50 per cent agreement (correlation) with both. There are formulae to work out the expected height, based on the average of both parents' height, that work pretty well. Only very rarely is growth hormone used as treatment, because it carries other potential health risks. We have not yet identified the genes involved in height, but a likely one may be related to the female sex hormone, the oestrogen receptor gene on chromosome 6, which strangely is important in bone growth in men as well as in women.

Why our genes ensure that men are usually taller than women is unclear, but so-called 'sexual selection' may have a role. For mysterious reasons, women of all cultures choose taller men as mates and men prefer smaller women. This selection pressure maintains the sexes' difference. Being tall for a man has obvious initial advantages, improving status and attracting more women. It can have its downsides, however. Tall men live shorter lives than small ones. In longevity terms, small is beautiful. For men with genes for more modest stature there must be other hidden ways to compensate to improve status and power. It may be coincidence but the three most powerful and infamous men in the last two hundred years –

Napoleon, Stalin and Hitler – were all under 5 feet 4 inches
(162.5 cm).

The bed-wetter

*At age eight J still wet his bed at night – unlike his friends and
siblings. His slightly anxious middle-class parents were
worried about any hidden psychological trauma or medical
problems he might have. After a year of talking and various
bribing and punishment incentives, they took him to the
doctor. He told them to be patient and gave them some
electric buzzers for the mattress to try. This didn't work and
they insisted on seeing a specialist in the hope of finding a
medical or psychological reason. The expert ran a few simple
tests to reassure them and took a family history. Both J's
uncles on both sides of the family had the same problem,
which resolved itself in time. J's nocturnal problems slowly
improved naturally without any treatment, although, even as
an adult, he still had the occasional 'accident', usually after
heavy drinking.*

Bed-wetting (the medical term is *enuresis*) is usually harmless.
In most cases, despite what older medical texts and folklore
have us believe, the cause is decidedly not stress or
psychological trauma: it is highly genetic. It is much more
common than most people believe, affecting 5–10 per cent of
ten-year-olds and 1 per cent of eighteen-year-olds. Research
has shown that it is usually inherited and you can almost
always find another affected relative. If a single parent was a
bed-wetter, nearly half of their children will be; If both parents
were bed-wetters, three-quarters will be.

Researchers have identified an association between bed-
wetting and three areas of the gene named ENUR1 and
ENUR2 on the chromosomes 13 and 12 and ENUR3 on
chromosome 22. These genes affect whether children will
need to urinate at night or how easily they can wake up when

their bladders are full. This is related to a lack of neural maturity and is somehow linked to problems of circadian rhythm (24-hour clock) as certain brain chemicals (such as vasopressin) can reset the clock and help the problem. Recent research has suggested the involvement of another mechanism – bed-wetters may have smaller palates and may sleep deeply and snore, perhaps altering brain oxygen. This is likely to be genetic.

For reasons we do not fully understand, it's much more of a problem for boys. Later in life the reverse is true, with 50 per cent of postmenopausal women reporting some form of urinary incontinence (involuntary leaking of urine, called *stress* or *urge incontinence*). Although the risk increases with the number of childbirths, nuns and other child-free women still suffer and we've found in our twin survey that it is 50 per cent heritable.

For parents the lesson is – before worrying – research your full family history (even if it is embarrassing to ask Uncle Frank!). In this way you can reduce family and child anxiety and hopefully avoid the secondary psychological problems that are common. However, you should still seek help in stopping the problem earlier, as there are now many effective treatments.

Wheezes and dirt

F was a fairly healthy nine-year-old. She'd had the occasional attack of bronchitis when young and a bit of mild eczema behind her knees (dermatitis). On moving into the family's new home, F started sneezing and wheezing at night. Her mother was slightly obsessive about cleaning the house and keeping it free of dust and dirt. She had read that dust mites were a cause of allergies and used to vacuum the house three times a day. They tried homeopathic remedies and soya-milk substitutes without success. One night her breathing was so bad they had to take her to the local hospital. Her mother suffered from eczema as a child and did not develop mild asthma until middle age, and she had no problems in the new house. Her father suffered only mildly from hay fever. One factor making F's asthma worse was found to be the new cat. F eventually improved with time and got used to the cat – but could never stroke it. By thirteen she led a fairly normal life but always had to take her inhalers with her.

Asthma and allergies such as eczema and hay fever are caused by a combination of genes and environment. Although there are hundreds of theories to explain asthma – from pollution to vaccinations to diet – the shared family environment appears relatively unimportant compared with the underlying inherited component and external environmental effects. The genes for asthma, hay fever and eczema overlap, so, if you or your parents are susceptible to one, you are likely to be at risk of the others. This risk will depend on your being in the 'wrong' individual environment at crucial times of life.

Allergies can be very specific for each individual. Having an allergic environment doesn't on its own cause asthma, but, once you have developed it, such an environment can make it worse in the short term. Asthma affects about one in five children mildly and about one in twenty severely. Rates seem to be doubling every twenty years in Western countries and even faster in places such as Greenland, because of rapid urbanisation – suggesting a change in the way we react to our environment. Allergies to cats and dogs are quite common. These may stem from the fact that we've started to live close to them only in relatively recent human history.

Asthma is essentially a protective response of the body against infections. Unfortunately, the response gets a bit carried away in certain environments that the body recognises as foreign or unusual. Among our ancestors, having the genes must have been helpful, and kept a few people alive against infections, particularly worm and parasite infestations, which were very common in their time. After a decade of failed genetic studies, several genes responsible for asthma have now been isolated. These include the ADAM gene on chromosome 20, which controls the repair mechanisms of the lungs when damaged, as well as immune-response genes on chromosome 5, which alter the way we react or overreact to infections.

A fairly convincing current theory for the persistence of allergy genes is this. While asthma and atopic (allergic) genes were perfect for us in combating worms and parasites when we lived in dirty environments, in our current ultra-clean, oversterile environments, babies and infants are deprived of our normal responses. We therefore later develop asthma, eczema and hay fever. Studies showing that babies who get many early colds and chest infections may be protected from allergy later in life add support to this theory. Living on farms gives babies early exposure to allergic substances (such as lipo-polysaccharides) and recent genes implicated in asthma such as TLR4 may modify our response to this. Perhaps giving our kids the early experience of a prehistoric rural playground with no vacuum cleaners or cleaning agents would do the trick and prevent allergies in later life.

Milk and good allergies

A was a normal, healthy, slightly overweight toddler, but as he grew older he became thinner and often reacted badly to his food. This became worse and he eventually had persistent diarrhoea, a bloated stomach and cramps. He was obviously now quite sick. His parents consulted their doctor and eventually a paediatrician. He was diagnosed as having genetic milk intolerance. The parents were surprised, as he had always liked milk and dairy products. When milk was removed from his diet he improved rapidly and was later able to eat cheese and yoghurts without further problems. He never drank milk again.

Intolerance to drinking milk, or, more accurately, an inability to digest it properly after infancy, is surprisingly common. It is actually the norm in most parts of the world, except in people from northwestern Europe, where only about 30 per cent are unable to drink milk. All children, wherever they are born, have the gene that makes the enzyme that is needed to digest milk protein (lactase). Among our distant ancestors (and other mammals), the gene was no longer needed after they finished breast-feeding and started eating on their own.

Approximately ten thousand years ago wild aurochs (a recently extinct type of cattle) were tamed simultaneously in the Fertile Crescent in the Middle East and in India, and domestication of the new cow or oxen species began. Dairy farming took off in a big way, particularly in certain grassy northern areas. Initially used for the meat, cows demonstrated other benefits, such as pulling ploughs and producing milk. A mutation in the gene continuing lactase production must have been very useful, and spread, allowing those people access to an extra source of protein and strength without expending energy hunting it.

People without the mutated gene were still able to benefit from goats and cows by eating cheese, letting the bacteria that live on cheese do the work of lactase. This may explain why people of southern Europe consume more cheese than milk, because the mutated gene is less common than in the north.

Food allergies, which are different from 'intolerance', are very common and caused by the body's formation of antibodies as a reaction to the food as a dangerous foreign substance. These allergies arose from humans eating new (in evolutionary terms) foods, domesticated by man in the last ten thousand years. The main eight substances (allergens) are peanuts, soybeans, wheat, tree nuts, eggs, fish, shellfish and milk.

Peanut allergy is particularly serious. Around 1 per cent of the population has it and rates appear to be increasing in Western countries. It affects young children and can stay with them through life. It is often life threatening – induced by just the smallest amount of the substance, such as a mother's peanut-butter kiss. Allergies to soybean are milder but more common – but, as over two-thirds of supermarket-sold foods contain some, it may be impossible to avoid. A number of companies are using genetic modification (GM) techniques to try to neutralise the genes responsible. Scientists have been able to deactivate the genes that switch on the production of the dangerous protein p34, which causes most cases of soybean allergy, and are only a few years away from wide-scale production. Other 'new foods' such as Kiwi fruit are being increasingly recognised as modern causes of food allergy.

Allergies to 'old' foods such as fish and shellfish, which have presumably changed little over the years, are more difficult to explain. They may be a consequence of our ancestors' lack of exposure to rare foods from outside their area for thousands of years. Some studies suggest that some early humans living by lakes or on the coast strangely did not eat fish, although it must have been in abundance. So, if you are allergic to lobster, the chances are you had ancestors who lived nowhere near Maine or the west coast of Scotland.

The difficult kid

M was always different from his brothers and sisters. More restless, and a poor sleeper, he lacked concentration and found it difficult to socialise with other children. When he started exhibiting problems at school, his parents took him to child psychologists and dieticians, but nothing much helped. His mum felt guilty, as she couldn't control his behaviour and felt it was in some way her fault. They learned that his uncle and grandfather had identical problems at school. M's problems resolved sufficiently with time and a short period of medication. He started to socialise better by the age of thirteen and, although he was hard work to live with, he found he was quite popular with girls.

Attention-deficit hyperactivity disorder (ADHD) is a condition characterised by severe hyperactivity, impulsiveness and inattention, with an onset before the age of seven. It affects between 2 and 6 per cent of children, predominantly boys, often being related to other reading or antisocial problems. It can exist in either very mild or very severe forms and can persist into adulthood with an increased risk of antisocial behaviour and drug abuse.

Although there is some evidence that extreme family and educational traumas can have an effect, ADHD is not usually related to upbringing or family environment. At least twelve twin and adoption studies have now shown it to be a strongly genetic trait with a heritability of over 70 per cent. The risk to children is about fivefold if one of their parents or older siblings is affected.

The origin of this genetic condition is believed to lie in the levels of the brain transmitter chemical dopamine, which acts as one of the brain reward mechanisms. Drugs that increase dopamine levels, such as Ritalin and amphetamines, usually help affected children. One of the main genes involved is called DRD4 (in particular a longer variant called 7R). The gene controls an enzyme called monoamine oxidase, which influences brain dopamine levels. Some cases need and benefit from medication but many are helped by parental and teachers' understanding. Eventually, there will be a reduction in family tension and a gradual easing of problems over time.

There were probably evolutionary advantages for adults who possessed this gene that explain its continued survival. The DRD4 gene is common in successful migrant groups and some South American Indian societies dominated by aggressive males good at hunting. The gene benefits the individual either by allowing him to be more forceful or by making him more attractive to females through displays of wild, novel or amusing antics. Alternatively, his restless nature makes him a successful wandering migrant, where in new lands he could find opportunities for acquiring resources and mating.

There are many famous adults who survived ADHD in childhood and perhaps benefited from it, including Buzz Aldrin, the astronaut who went to the moon – the ultimate migrant. There is also some evidence from studies of primitive African tribes that children who had these traits survived long droughts and famine better than other, 'easy', children. This may be because they demanded much more attention from the mother, while the 'easy kids' quietly faded away.

The geek gene

H and S married late in life. Neither of them had had much time for socialising in the fast-moving computer and biotech industries. They were nonetheless delighted when they found out they were due to have a baby boy. All was well initially and he started to walk and say a few words just after his first birthday. At the age of two they noticed that he wasn't using any new words, and didn't interact well with other kids. By the age of three his language regressed. He refused to play with adults or children and didn't make eye contact. A paediatrician diagnosed him as having a mild form of autism called Asperger's syndrome. H and S read the literature on the subject and found that many experts believed this was partially due to a lack of parental warmth and bonding. Besides being devastated by the diagnosis, they felt far worse thinking their behaviour may have caused the condition.

Autism is described as the failure to develop social relationships, characterised by a total lack of interest in the social use of language and an inability to perceive social clues and contexts – in other words, to see other people's viewpoints (which is also known as having empathy). Autistic children tend to focus on objects rather than people. While many are diagnosed with developmental or learning problems, some children are of normal or above normal intelligence and can have special abilities. Although childhood autism at one time was considered rare, it probably affects one in two hundred children, mainly boys. Modern research supports the concept that many disruptive, childhood and adolescent behavioural disorders are part of a spectrum of behaviours that have a strong genetic component. These all have a number of genes in common that affect dopamine, serotonin and other brain chemicals, and are inherited from both parents.

Autism and its related milder form, Asperger's, are now more loosely defined as one of a range of autism spectrum disorders (ASD), whereby children all have social communication problems but variable intelligence and reading skills. Asperger's children are typically of very high intelligence and, although they always lack certain social skills and the ability to show empathy, they can lead fairly normal lives, becoming famous designers, scientists and in some cases doctors. These conditions were until recently blamed unfairly on the behaviour of parents (in this case mothers). Psychologists and doctors used to believe it was caused by cold and unaffectionate mothers ('refrigerator mums').

Recent research from a large number of twin studies has shown this theory to be nonsense. Autism disorders are around 90 per cent heritable, with the concordance (agreement) in identical twins being 65 per cent – leaving little room in most cases for an important influence of family environment, parenting or immunisation. The genes causing ASD are close to being identified. Areas on chromosomes 7 and 15 have now been implicated that contain one of the key genes (such as WNT2 or RELN), important in brain development, or the GABA receptor gene, important for calming down brain signals. Another is the AUPR gene.

Recent surveys have suggested that ASD is on the increase, particularly in areas such as Silicon Valley, California, and areas near Cambridge, England, home to many hi-tech electronics and biotech industries. One hypothesis to explain this is that the special advanced mental skills and genes needed to be a highly specialised scientist or an electronics boffin (commonly referred to as a computer geek) have a side effect of causing mild and usually undetected autism with resulting poor social and communication skills. These adults are likely to carry the important forms of the Asperger's genes. When a very brainy 'geeky' male meets and mates with a brainy female, the dose of genes from both parents may be such that a child with Asperger's may often be the result, particularly if it's a boy.

The fact that very intelligent children and adults often lack social skills is not a surprise to most of us. That genes are responsible for this strange combination is more interesting. The reason males are particularly affected is a puzzle and may be due to Asperger's being an exaggeration of the normal masculinisation of the brain as well as the inbuilt advantage females have for social skills and communication. This advantage may come from the X chromosome, which females have two of compared with a single copy in males. This suggests that, for women, as we discussed earlier, genes on the X chromosome are a calming influence.

Nevertheless, in evolutionary terms, for certain men, being extra brainy seems to have an advantage, even at the expense of 'geekiness'. It may be unfair to label these 'gifted' individuals as having a disease – otherwise, by the same token, we should also call successful women with extra social (empathy) skills who are unable to read a map or program a video 'diseased'.

The slow reader

H was six and struggling at school. Unlike his classmates and his sister before him, he couldn't read properly. His parents were both professionals and felt they had a good home environment. His mum, who worked part time, felt guilty about not giving him enough attention and reading time. She had constant fights with her husband for working too late. His six cousins were all Italian and read well at the same age. H's parents worried they should be doing more to help him. A friend suggested he may have dyslexia, but they were afraid to send him for formal testing for fear he may be stigmatised.

Reading ability and achievement of reading milestones in otherwise normal infants and children is highly genetic. This has been clearly shown in family, twin and adoption studies. Any deficit in language ability occurring in someone with normal intelligence is now known as specific language impairment (SLI). This term has replaced the obsolete terms 'dyslexia' and 'word blindness'. Affected children, as well as reading poorly, have trouble pronouncing complex words and repeating nonsense words. Studies have indicated they use a different information-processing pathway in response to reading compared with normal children, which involves using more guesswork. A key area seems to be the part of the brain that memorises sounds for a short period of time. The disorder is twice as common in English- and French-speaking countries as in those with much simpler phonetic languages, such as Italy and Spain.

Several genes linked to these SLI problems have been implicated recently on chromosomes 1, 6 and 15 – but other areas are under investigation. There is now believed to be a large genetic overlap between autistic disorders and specific language impairment, because one in four parents of autistic children had problems reading as a child. Even the correct use of grammar in speech is known to involve a gene, located in the same area as some of the genes for autism on chromosome 7q. This gene controls a grammar centre of the brain, which, when defective, has disastrous social results. In one very rare but large British pedigree, known as the KE family, affected family members are unable to communicate properly or understand long sentences, despite being able to clearly understand individual words. They are also unable to understand or use the past tense. The gene FOX2P on chromosome 7q has recently been identified as the likely culprit for the grammar problems in this family – but it is unlikely to explain most common language problems that will involve other combinations of genes.

How infant brains acquire speech and language is slowly becoming clearer. Babies are born with an inbuilt system that allows them to recognise many different sounds from all possible human languages, be they English, Finnish or Japanese. If they don't hear these sounds within a few months, these specific brain cells die off, and cells expand in other areas. Building blocks for a universal human grammar are also probably present from birth in all babies using a so-called 'language-acquisition device'. This is an innate capacity for syntax that prepares them to build a language *de novo* – from the merest scraps of linguistic input. Language probably evolved in the form it has, so that young children, who prefer learning in chunks, could learn it easily. But picking up a new language is much harder for adults, who prefer learning in smaller, more discrete pieces. Such is the flexibility of the infant brain that babies can process and familiarise themselves with new words and sounds even as they sleep. Studies have also shown that parents automatically adapt to the needs of their children. Within a few weeks, babies without prompting start babbling baby sounds, which are not random but careful practice sessions with their mothers involving the language centre in the right side of the brain.

Communication skills are generally much slower to develop in boys than girls. This difference in the starting points may be to some extent an evolutionary phenomenon, as language was more crucial to our female ancestors. Pleistocene females spent most of their lives socialising with groups of other women for the purposes of food gathering and mutual protection. They needed good communication skills to tell each other exactly where the best fruit could be found and to maintain their hierarchy. Men hunted alone or in much smaller numbers. Areas of the brain dealing with language in the left temporal lobe appear to differ in adult men and women as a result. Why genes for reading problems have evolved or persisted is hard to explain, although in prehistoric man, other than reading footprints, it was unlikely to have been a major handicap. The good news is that it is not the parents' fault and the vast majority of kids grow out of it. There are many famous 'dyslexics' and slow readers to substantiate this, including Albert Einstein, who couldn't read until he was seven – but managed to catch up.

Are there genes for intelligence?

N's parents were disappointed at the lack of progress he was making at school. They had high ambitions for their son. He seemed to be working hard, doing his homework and appeared to like the extra tutoring he received from his parents. Because of his poor marks he was given an educational assessment with a IQ test, which showed him to be slightly below average for his age. They were devastated and felt they had let him down badly – perhaps they had over-worked him or sent him to the wrong school.

Having difficulty learning to read is not a strong indicator of mental ability or intelligence. But is intelligence something that is acquired, or is it, too, genetically determined? This subject has been understandably controversial since the first studies were performed in the 1860s by an early Victorian African explorer turned geneticist, Francis Galton, a cousin of Charles Darwin. He showed that very bright people (geniuses) were more likely to be closely related to other bright people than could be down to chance. Critics reacted strongly and suggested it was heretical to suggest that all men were not created equal. This led to dogmatic claims by religious leaders and so-called behaviourists (such as J B Watson) in the early and mid-twenty centuries that, given the right education and environment, any child could be moulded into any profession (doctor, lawyer, artist, merchant chief, beggarman or thief) regardless of talent or the background of his or her ancestors.

Intelligence is one of the most studied human traits. Many hundreds of subsequent twin, family and adoption studies involving more than 35,000 subjects have all consistently shown heritability of measures of intelligence. On average the studies show that identical twins (with the same genes) have around 85 per cent similarities in IQ scores (conventional tests of 'intelligence'), which is close to that of the same person tested twice. Even when adopted identical twins reared apart in different families are tested, they are about 79 per cent similar. Results for pairs of brothers or sisters or fraternal (nonidentical) twins, sharing half their genes, display about 50 per cent similarity, compared with correlations of under 10 per cent among adopted children or unrelated individuals. Overall, these different studies give an estimate that 70 per cent of 'intelligence' is due to genetic factors. Strangely, the innate genetic influence seems to become clearer and stronger with age: tests done on young infants show a lesser effect by genes (less than 40 per cent and proportionally more environment), compared with studies of the elderly, where an even greater proportion (greater than 80 per cent) is genetic.

It should be pointed out that IQ is a crude test of intelligence that averages out the scores of a number of different brain abilities. However, when you perform a wide range of brain tests, such as short- and long-term memory, reflexes, visual recognition, speed of reasoning and so forth, they all agree (correlate) with each other to some extent. This means someone who is good at one test is likely to score well (but not the same) on a different test of a different skill. Indeed the same genes appear to influence all the different skills. This suggests that we inherit some central 'intelligence genes' (sometimes referred to as 'g'), which give us different processing speeds and rates of learning, allowing us to expand our brains in different ways.

Recent twin studies have shown that the amount of grey matter contained in the brain may be one of the ways genes affect us. The ASPM gene has been found in animals to control the number of brain cells (neurons) that eventually develop – defects in the gene have been found to cause very small brains in a few rare human families – suggesting some potential role in intelligence. But we shouldn't overemphasise the cleverness of genes in isolation. Genes make simple proteins and the blueprint for the brain. Once formed, the brain takes over its management, makes its own memories, reacts to experience and has its own variable desire to learn and expand – and can probably increase or decrease the influence of genes to achieve this. Saying that IQ is predominantly genetic shouldn't produce a fatalistic attitude, as education and family input are also important. IQs have been steadily increasing in many countries; many of the slight discrepancies between groups disappear when the more socially deprived kids improve their status. Japanese children have improved their average IQs by 20 per cent over the last 25 years through better schooling and education. An important recent study of 640 seven-year-old twins in the US has shown that in deprived and impoverished kids, IQ is influenced predominantly (60 per cent) by family and social environment. In contrast, in well-off kids with good homes and school situations there is no discernable effect of environment and all the influences on IQ are genetic. Improving social and educational factors in deprived kids can therefore have an important effect on IQ, despite its strong genetic influence.

While there are several well-known families of geniuses (the Darwins, the Bachs, the Bernoullis, the Mozarts), they were all financially well off and well educated. Every so often a genius is born who appears to be created out of the blue. A number of geniuses, such as William Shakespeare, the mathematicians Carl Gauss and Srinivasa Ramanujan and the inventor Michael Faraday, were exceptional in that they came from unknown, uneducated and poor parents. None of their brothers or sisters or offspring showed any exceptional gifts, showing that genius arises only when the exact configuration of genes is put together by chance – which is rare.

It is still an evolutionary mystery that, compared with apes, our brains grew so quickly to be three times the size, and our human brain tissue has many times more gene activity. Ideas to explain this range from sudden climate changes or food shortages that caused the sudden need for greater reasoning and tool making for survival to the theory that females found brainy males 'sexy'.

4

Genes and the Terrible Teens

The teenage and adolescent years are usually remembered as the most traumatic and insecure time of our lives. For parents they're also the most stressful. Most parents and adolescents get through the experience without permanent harm, but the role of our genes can be crucial in many aspects.

Why is life so unfair?

S had been a delightful cute kid, loving, expressive and fun to be with. Suddenly it all changed. On turning thirteen he became moody, sulky and rude. He would often react badly to conversations at the table, mutter a swear word under his breath and leave for his room, slamming the door without finishing his food. Despite not speaking to his parents, S spent many long hours on the phone or on his Internet chat service talking to friends. He always denied anything had changed and after rows about phone bills complained bitterly to friends that he hated his parents, who he said didn't understand him. Why was life so unfair?

Research has recently uncovered the fact that part of a teenager's seemingly unsocial behaviour is due to the failure to efficiently process signals from other humans. The size of the brain has by now reached the adult maximum of

1.4 kilograms, but the frontal area of the brain (the so-called higher centres responsible for planning, language and social interactions) are still very much under development. This means that these brain areas are exposed to a wide range of subtle signals from teachers, friends and parents and yet they can't correctly process all the information. In other words, teenagers are constantly misinterpreting other people, and in the confusion overreact and act defensively to hide their mistakes. This is part of the normal process of brain development and may be seen as cerebral 'growing pains'.

Genes play a role in this by varying the speed at which the brain changes – each of our 100 billion brain cells (neurons) is controlled by genes that tell it how to respond or develop. The children with fast changes over a short time have the most problems, whereas others who go through slow steady changes are able to cope more easily with their surroundings. Teenage years are a time of major mood swings, and this is more marked in females – probably because hormone levels change faster or earlier. In later life, mood swings are an adaptive part of female behaviour that may have evolved partly (according to some male theories) to test male resolve and commitment.

The teenage years are exclusively a human phenomenon. Other mammals pass quickly from total dependence to independence – even apes don't remain teenagers for long. We may have evolved this transient period, as our brains needed more time than those of other animals before independence and reproduction to develop crucial skills for later life. These would include direction and food finding, hunting and making tools, socialising and communicating with peers, and of course experimenting with sex. The teenage years are a way of practising these skills by pushing the boundaries as far as they can in a relatively safe environment. Universally, teenagers will drive their parents to distraction – but rarely get murdered by them.

Budding too early?

A was the perfect daughter, accomplished at school, tidy, sociable and pleasant. When she was eleven and a half years old her body shape began changing rapidly, with breasts and pubic hair appearing within a few months. Her personality changed also, and at twelve all her interests and energies focused on boys. She became difficult and uncommunicative with the rest of the family. When she was thirteen, her parents overheard her talking on her mobile phone about wanting to have sex with a boy at school. Had they encouraged this – or was the school responsible?

We have observed how much teenage behaviour is explained as practising adult courtship skills, taking risks and testing boundaries of adult authority. Conflicts with parents arise, as the physical stage of development is unfortunately not usually in step with the mental or social skills of the teenager. Genes influence the timing of puberty and much of teenage sexual behaviour, which often varies markedly between siblings brought up together. Puberty and age of sexual interest are mainly triggered by body weight, which switches on the necessary genes. As women have a limited fertility period, evolution (and therefore current genes) favours women with early sexual development – provided the body is healthy enough and has reached the correct size.

African-Americans and Afro-Caribbeans have on average earlier puberty and onset of periods (menses) than Europeans. The average body weight that triggers periods in girls is around 105 pounds (48 kg) – the point at which the body usually has sufficient nutritional reserves for a successful pregnancy. The fat cells signal the brain with a natural chemical substance called leptin, which triggers brain production of hormones that stimulate the dormant ovaries (eggs) to start working. In boys, puberty occurs slightly later and is triggered at a slightly higher weight.

The average age at which a girl's periods start has dropped steadily over the last hundred years and is now between twelve and thirteen years, (with a wide range from nine to sixteen years. This is due to better nutrition and greater numbers of overweight children – a trend that is steadily increasing. Daughters will tend to have slightly earlier periods than their mothers. Twin studies have shown that the timing of the first period is around 40 per cent heritable, and the genes responsible are unrelated to genes controlling when periods stop.

Among our ancestors, and in primitive populations today, periods were triggered only when a girl was mentally and physically healthy and sufficiently well nourished. This probably sent signals to our male ancestors indicating that the teenager was a good bet – being fat enough to have had a good family support network to help her obtain food, who would also be around to support a baby. These girls usually started menstruating at around fifteen or sixteen years of age and often were not fully fertile until they were twenty years old. The modern fast-food diet has messed up this safety mechanism and many girls with burger-induced early puberty (due to high fat and calorie intake and possibly animal hormones) are the ones least able to cope mentally and socially with childbirth.

As periods are linked to health, fertility and availability for sex, age at first intercourse is also related to sexual maturity. Twin studies have shown that the age of first sexual intercourse is also partly under genetic influence. Just as the age of puberty is earlier, ages at which girls lose their virginity have been decreasing progressively. Currently in the UK, 25 per cent of girls and 30 per cent of boys have had sex before they are sixteen. In the US in 1950 only 7 per cent of sixteen-year-olds had had sex, but by 1982 the percentage increased to 44 per cent. In most Western countries (US and most of Europe) the decreasing age has started levelling off since the 1980s. In the UK, the rates of underage sex have been static since the 1990s.

As well as genes, culture and family environment also play a role in the age of first intercourse. Children who are abused or who have unsettled or disturbed childhoods without a father tend to have earlier and often unhappy sexual relations. This may be related to these children's understandably greater insecurity about their longer-term (and subconsciously genetic) future. Recent surveys have shown, however, that poor educational level is an even more important factor than family environment in relationship to early sex and high risk of pregnancy. The vicious circle whereby teenage mothers produce daughters who repeat the same behaviour is thus a complex mixture of genes and social and educational environment. Large family-based psychology studies from the US have shown surprising effects of parental influence on children's attitudes to sexual freedom. Children reacted in opposite directions depending on their sex. A daughter was more likely to mirror her mother's attitudes and take the opposite view to father – and vice versa for boys who took after their fathers.

Among children who are well educated there is no clear link between early puberty or early sex and any special emotional or sexual problems. However, two-thirds of girls who reported having sex before they were fifteen, were not emotionally ready for it and regretted the experience. In countries with open attitudes towards sex education (such as the Netherlands and Sweden), age of first intercourse is often delayed and the risk of teenage pregnancies is ten times less than in the UK or US, where a girl who drops out of school at sixteen has a one-in-three chance of having a teenage pregnancy. The message from this is that you can't alter your kids' in-built interests in sex – but, with education, you can make it a safer experience.

Spots and chocolate

When D was fourteen and starting to flirt with boys, spots and acne were becoming a major problem – and wouldn't go away. She went on special diets, in particular avoiding sugary items and chocolate (which she loved). These sacrifices didn't help and made her more frustrated and unhappy. Her mum read magazines saying spots were due to poor hygiene and diet and she bought vast amounts of expensive facial cleansing agents. Despite going on more and more bizarre diets, taking Chinese herbal remedies and spending hours in the bathroom, nothing seemed to work, and D started to get depressed and became progressively socially isolated.

Acne is endemic in developed countries, affecting around 8 million Britons and 50 million Americans. Around 90 per cent of teenagers have some spots – so it could be considered a normal part of puberty. However, around one in six teenagers is affected so badly that it may lead to permanent scarring – and it is one of the commonest traumas of adolescence, leading to anxiety, personality disorders and occasionally suicide. The reason it is not a trivial problem is that we are all genetically programmed to admire a good healthy complexion and be repulsed by a flawed one. This is probably a throwback to our ancestors, whose survival used to depend on their ability to detect early infection or infestation in a stranger by looking at their skin. This explains the dramatic transforming effect a single spot can have, even on the most stunning face.

Despite popular opinion and neurosis-inducing commercials, acne in most cases is not related to poor skin cleansing. Some cultures equate spots with having too much pent-up sexual energy and hormones. Twin studies we performed in more than two thousand twins for acne have shown it is over 80 per cent genetic at all levels of severity. It is related to over-secretion by glands in the skin that produce an oily substance called sebum. In teenagers, sebum glands are working overtime due to the effect of high levels of sex hormones that turn on the genes that make more sebum. It's also related to abnormal shedding of the skin around the hair follicles, and bacteria such as *Propionibacterium acnes* move into the space around the follicle and cause local irritation. Genetic resistance to these bacteria could be another factor in susceptibility.

It appears that acne is a recent human phenomenon. Surveys have found much lower levels of acne in developing countries such as Brazil and India and among the Bantu of South Africa, with even lower rates in those living in rural rather than urban environments. A recent study looked at rates in two primitive populations, the Aché of Paraguay and the Kitavans of New Guinea. More than 1,300 people were closely examined and not a single spot was seen or reported.

The explanation could be diet. Hunter-gatherer diets are very different from our own. They have low fat, but, more importantly, virtually no starch and complex carbohydrates such as those found in potatoes, bread and pasta, which increase our blood sugar levels, and lead to other diseases such as diabetes and obesity. These carbohydrates now make up a substantial part of our diet but have been around for only the last few hundred years or so. So acne is primarily a genetic disease that may be exposed or exacerbated by our Western diet. This suggests that having these genes may have been of use to some of our ancestors, for whom a greasy skin made it healthier and less liable to ageing, drying and sun damage. Whether carbohydrate-exclusion diets work in treating existing acne is unknown and the regime would be pretty difficult to maintain in the long term.

Nowadays, many effective treatments such as antibiotics, anti-androgens and vitamin-D-like drugs exist to reduce the potential long-term effects and trauma of acne. For those of us in countries exposed to starchy Western diets it's clearly a genetic disease; for the rest of the world, having a spot could be perceived as a sign of success and affluence.

Early habits

E was thirteen when his mum caught him with an empty cigarette packet in his bag. His mum was furious and partly blamed her husband, who still smoked more than twenty cigarettes per day. She herself had smoked from fourteen until the age of twenty-five, when she managed to give up with great difficulty – because a boyfriend didn't like the smell. She was now very antismoking and was determined her children would not get addicted. She was worried that smoking cigarettes was the first step towards harder drugs and other problems. She wondered whether the fact that her husband still smoked was likely to be a major influence on her son.

Experimentation with smoking, alcohol or drugs is becoming part of a modern teenage ritual, which is on the increase in most countries, and associated with later addiction problems. Despite the obvious cultural influences, such experimentation appears to be moderately genetic. The number of children under fifteen who try cigarettes varies widely from 5–30 per cent depending on area and country. A large study of Dutch teenage twins showed a heritability of 39 per cent for onset of smoking. The effects of family education, culture and the influence of friends, and school are more important than genes. Studies have shown no significant effect of current parental smoking behaviour on children, once these other factors and genes have been accounted for. A recent New England survey of five thousand children actually showed a relationship between the number of films they had seen recently depicting smoking and whether they had recently taken it up themselves.

The longer teenagers smoke, the greater appears to be the effect of genes. Whether or not they become addicted in the long term is highly genetic, as we see later – but in teenagers addiction is difficult and dangerous to predict. Some recent studies have shown that a small proportion of genetically predisposed children may become addicted after smoking only two or three cigarettes. In most kids, addiction occurs after around six months in boys, but after only three weeks in young girls aged twelve to thirteen, who appear much more susceptible.

One-third of fifteen-year-old children in the UK have already experimented with cannabis and 7 per cent with stronger drugs. Levels of use are even higher in the US, with around 50 per cent of kids having tried cannabis by the age of eighteen. Twin studies of cannabis use in Australia and the US among adolescents have shown clear genetic influences. These get stronger the more regular the use (with heritabilities of up to 80 per cent). The abuse of dangerous substances such as glue fumes, or harder drugs such as cocaine or heroin at a very early age, normally suggests social or psychological problems that may have been present for a while. Despite fast-moving changes in attitude and culture, levels of serious addiction have remained fairly static at around 1–3 per cent for the last thirty years – showing the importance of genes and a constant pool of susceptible people. Nevertheless, parents shouldn't be complacent. A recent Australian study found that children using cannabis before the age of seventeen were two to five times more likely to have drug-related problems later in life. This couldn't be accounted for by genes, parents, smoking, alcohol or mental state, and suggested the importance of the peer group and easier access to hard drugs.

Alcohol use has increased over the last four decades. One in five children under sixteen is now drinking at least once a week in the UK. Early drinking is a major risk factor for long-term addiction. The age at which kids begin drinking alcohol regularly is under a genetic influence, which is moderated by cultural background. One benefit of religion is that it may reduce drinking and smoking, even in genetically pre-disposed individuals.

How genes influence early experimentation with cigarettes and alcohol is speculative. One theory may be related to taste genes. Many sensitive children cannot tolerate the bitter tastes of beer and other alcoholic beverages; others lack the gene completely and have no problems. Although we slowly grow to like bitter tastes, genetically determined taste receptors may put off many kids early. Another explanation is chemical rewards. Young people recently attending a medical clinic who reported abusing illegal drugs were four times more likely to have two copies of an unusual form of a gene (FAAH) than people without any drug or alcohol problems. This gene provides the code for an enzyme responsible for deactivating natural cannabinoids, which act on the same brain receptors as the psychoactive component of marijuana. The enzyme is involved in reward and addiction pathways in the brain. The gene variant causes a build-up of natural cannabinoids, so that abusers need to take much more of the drug to achieve an effect and so become addicted.

Finally, genes for risk-taking behaviour may also be involved in addictive behaviours. Some of these genes may not be all bad in the long run: Richard Wagner was a serious teenage gambler and Graham Greene was fond of Russian roulette as an adolescent.

Skin and bone

J's mum N was worried. Her thirteen-year-old daughter, who was of normal height and weight for her age, had started to go on special diets, as suggested by friends. Her weight was already starting to fluctuate from month to month and her periods, which had recently started, had stopped again. After evading questions for a while, J was eventually cornered and admitted to feeling overweight with chubby cheeks and 'fat bits'. J was worried that, if she didn't do something about it now, she would lose out in a big way later. Her mum, now in her late thirties and divorced, had experienced a difficult and sometimes unhappy childhood and had suffered from anorexia nervosa when she was an adolescent, which required medical help. N believed she had been a devoted and supporting parent – but maybe she hadn't, and this was her daughter's cry for help.

Eating disorders such as anorexia (undereating) and bulimia (cyclical overeating, vomiting and undereating) affect around

10 per cent of adolescent girls to some extent and 3 per cent severely. Eating disorders exist, but occur much less frequently in boys. In countries such as the US, mild eating disorders and obsessions with diets are reaching epidemic proportions, with surveys showing 40 per cent of normal-weight fourteen-year-old girls believing they are overweight and need to diet. The first reported case of anorexia was only three hundred years ago – so it was unlikely to have been a problem for our ancestors, for whom looking overweight was a major advantage. Being chubby was previously a sign of health, fertility and success – the fatter the young bride, the more the prestige. Some early Islamic countries had fattening rooms, where a bride-to-be could gain weight for a few weeks – a far cry from today's 'health' farms and spas.

The causes of eating disorders in teenagers are a mixture of culture and genes. Twin studies have shown that the most common problem, bulimia is considered moderately genetic (40–50 per cent), and anorexia strongly genetic (70–75 per cent). If the child also suffers from a tendency towards anxiety or depression (also highly genetic), her risk increases even further. Any child who early on in life displays an interest in serious dieting increases her risk of having permanent eating problems tenfold. Daughters of mothers who had eating disorders have about a tenfold increased risk of developing the same condition. Eating disorders involve a false perception of body image with girls overestimating their shape. A large twin study looked at whether perceptions of body image were predominantly genetic or environmental. It found that the ability of girls to perceive actual weight and shape was at least 50 per cent heritable, but that cultural influences predominantly determined their ideal desired figures.

A number of genes have been implicated as having a role in eating disorders. The serotonin receptor gene (a form of the 5–HT (2A) gene) involved in controlling levels of the brain chemical serotonin is one. The oestrogen receptor (type 2) gene and the potassium channel gene are others. So far the causal role of these genes has not been confirmed convincingly.

Environmental influences thought to be involved in eating disorders have been traumatic birth problems, teasing at school and marital problems in the parents. Studies have shown all these to be of very minor overall influence. In our scenario J's mum has every reason to be concerned – both because J started dieting at an early age and because of her own medical history. If the problem is serious, J needs professional help. If the problem is milder, encouraging exercise rather than dieting seems to be a much safer and effective way of dealing with concerns about body shape and image.

Off-key genes

S sat faithfully through regular piano and violin lessons for more than eight years – with disappointing results. Despite reasonable hand skills and above-average intelligence, she could not seem to excel and found it difficult to make progress. She told her parents she wanted to give up. Her parents couldn't understand, as they'd played music to her since birth. They felt they had provided a perfect musical environment – lessons, instruments and encouragement. They were particularly upset, as they never had the same chances in life, and their parents never gave them lessons or instruments. They thought she was deliberately trying to fail.

Pitch perception, which is the ability to hear whether a note is played in or out of tune relative to others, is now known to be 80 per cent genetic. This figure is based on a large twin study we performed using the responses to 24 well-known pieces of music, played in and out of tune. We found that one in twenty people were completely tone deaf and made random guesses in the tests, and one in four had poor pitch perception. Most people would score about 19–22 out of 24 and 10 per cent scored top marks. Despite her 'perfect musical environment', S has less than perfect pitch perception and will never make it to the top. While she may enjoy music,

S will find it difficult to compete with other music students. One or both of her parents probably also have faulty pitch perception – but like many people are unaware of it.

Using brain scans, scientists have discovered that certain areas of the brains of professional pianists are three times larger than those of nonplayers. This may occur because 'musical' genes allow the brain to expand itself in certain areas when exposed to learning. Recent functional brain-scan data show that, when people listen to complicated pieces of classical music, many different areas of the brain are involved, including those for memory, hearing, emotion, cognition and perception. An area called the *ventro-medial frontal cortex* in the front of the brain appears crucial for pitch perception – and in most people the right side of the brain does most of the work.

At birth we all have perfect absolute pitch (the ability to recognise and copy a single note perfectly without the need for a reference) but it unfortunately lasts only a short time. This ability is believed to be a stop-gap method to allow communication with our mothers, before the ability to detect relative pitches kicks in. At birth, babies are receptive to every conceivable sound that a human being can make. In other words, we are all born equally prepared to hear any language – Finnish, Korean or English – or any form of music. Depending on what we first hear, areas of our brain expand and other unused neurons wither away for ever.

Music is universal to all human cultures – although when humans first made music is debatable. The earliest musical instruments discovered so far are flutes (recorders) from 53,000 years ago. But we may have been creating music well before that. Amazing similarities have been noted between the music made by whales and humans. Although whales have a larger range, both species use similar rhythms, spacing and pitch, require considerable memory, and can teach others a new melody. Some Inuit tribes have the ability to hear and imitate whale songs. Bird songs are also very similar to ours in composition: for example, the canyon wren, which copies the opening of Chopin's *Revolutionary Etude* (or did he copy it first?). Whales and humans diverged in evolutionary terms over 60 million years ago, and yet our musical abilities are strikingly similar, suggesting the genes controlling this part of our brains may be very ancient indeed.

In evolutionary terms, musical ability may have developed in animals as a signal of high mental fitness in a potential mate – and subsequently became an attractive quality sought by females. Many male animals use their voices and songs to attract female suitors – such as numerous birds and the natterjack toads, which are the planet's best croakers and can be heard by females over a mile away. Fish (gobies) who live on the bottom of rivers have to pitch their love song in the 80–200 hz range or they go unnoticed. Romance could be the reason why humans have developed different vocal pitches.

Musical ability obviously isn't just about pitch perception: it also takes an innate sense of rhythm as well as the specialised motor and coordination skills to use an instrument or voice to the highest level. As now seen regularly in karaoke bars, many of us, helped by alcohol and peer pressure, have the confidence and skill to sing – but lack the key circuits in the brain and ear to tell us whether it's horribly out of tune. Perhaps the fact that music is such a basic and ancient part of our human makeup makes listening to it (whether good or bad) such an intense and emotional experience.

Lazy genes

T was definitely lazy. He was fifteen years old and conserved as much energy as possible. He spent many happy hours every day on the couch in front of the TV, slept as much as possible, and did the minimum amount of schoolwork. He hated walking and hated sports even more. While his brothers always performed well at football and other sports, T had no obvious skills. He couldn't throw a ball far or accurately, couldn't play tennis and performed badly at field sports. His lethargy and lack of sporting prowess was often a cause for his being bullied at school.

Most sport participation, motivation and skill is genetically determined when individuals are young, particularly for ball and athletic sports. Studies of twins and their parents have shown that willingness to play sports and exercise has a genetic component. In our twin study, we asked women how many hours of sport or exercise they spent outside of work per week. A study of 924 twins explored their theoretical likelihood of participation in 120 different leisure-time activities if time, money and health were not an issue. These included going on safari, extra education, night-clubbing, volunteer work and risky pastimes such as surfing and hang-gliding. How often the twins said they would do these activities was over 50 per cent heritable – an indication of motivation and laziness.

Sport is genetically a way for humans to display their variability in fitness as potential mates – showing off their 'good genes'. It is important to recognise that some children do find it harder physically and mentally to exercise than others. This is because of their genetic makeup. In contrast with sporty children, when these predisposed individuals try to exercise they are getting much less chemical 'reward' of endorphins and dopamine – and therefore more pain than gain. These kids need other activities to stimulate their reward centres.

In 1984 a 252-pound (112-kg) self-confessed slob and heavy smoker, Canadian Peter Maher, accepted a wager from friends that he couldn't run a full 26-mile marathon. Against all the odds and to the shock of his friends he won his bet. He went on, after losing 140 pounds (62 kg), to be a great marathon runner – achieving times of within six minutes of the world record. Why was this man with an inherent running and endurance ability initially so lazy? Laziness is a natural state. Most animals are naturally lazy – sloths, cats and lions are good examples. When mice are put on treadmills in labs they do anything to avoid jogging. The reason is a survival instinct – to conserve valuable energy stores before the next really important activity arrives. When our ancestors were not hunting or mating they probably spent a lot of time hanging around being idle. There was no point in using up calories that you'd spent so long building up. The trick to survival was to know when it was worth lifting your finger – when your life depended on it, for instance, or when it might increase your chances of catching a mate.

Teenagers are particularly skilled at this survival strategy. Studies of teenagers in hunter-gatherer societies with lifestyles similar to those of our ancestors (such as the !Kung of the Kalahari) are enlightening. They show the same work-shy traits, particularly the girls, who were observed to work less than three minutes every hour – much less than their younger or older sisters. The other members of the tribe complain about teenagers as much as we do, but attribute this to their need to conserve energy and build up their fat reserves before they can reach maximal fertility and have babies. So next time you see a lazy teenager, be more understanding, tell them to relax, put their feet up and build up their fat stores for tough times ahead.

Champion genes

L was fourteen and gifted at most sports, but found he particularly excelled at long-distance running. He was always the best in his school, usually beating older students. His mother always encouraged him. She had been a runner when younger and had competed for the state championships – but his father was more sedentary, preferring armchair sports and pool. There was a debate in the family as to how far to push the sporting aspect of his life to the possible detriment of his academic studies. They had been approached by the national junior sports academy to enrol L. His mother eventually convinced her husband and after years of training and dedication L went on to considerable national and international success.

The importance of genes in competitive sports has recently come to the fore. Some organisations, such as the Australian Institute of Sport, now choose young athletes for selected intensive training based on their family history of athletic success as well as their current performance. Twin studies in our unit have shown that generally there is only a weak genetic influence on participating in social sports – such as playing golf once a week. The genetic influence becomes much stronger the greater the level of success and competitiveness. A Finnish study put 742 previously sedentary volunteers through a twenty-week exercise regime. After training, some people showed hardly any improvement in their body's ability to use oxygen, while some showed a massive leap. Different physiological adaptations, they found, tend to run in families.

Anyone watching the Olympics can't help but notice the importance of genes in determining, on average, who wins which event. Most successful sprinters are well-built men and women of recent African origin and marathon runners tend to be small and slender and of European, Asian or North African origin. Champion swimmers are usually of European descent. These differences are not due to attitudes, environment and training but to their genetic makeup. Africans have more muscle and less fat than Europeans and Asians and are therefore better sprinters. Europeans and Asians perhaps evolved mutations in their genes to help them migrate and endure the long distances out of Africa into harsher climates.

Human muscles contain a genetically determined mixture of both slow and fast fibre types. On average, we have about 50 per cent slow and 50 per cent fast fibres in most of the muscles used for movement. The slow muscles contain more 'battery cells' (mitochondria and myoglobin), which make them more efficient at using oxygen. In this way, the slow-twitch fibres can fuel repeated and extended muscle contractions such as those required for endurance events such as a marathon. Fast-twitch fibres fire more rapidly. Having larger numbers of fast-twitch fibres is a major asset to a sprinter, for whom time is of the essence. The slow-twitch fibres, on the other hand, fire less rapidly, but can perform for a long time before they fatigue. Olympic athletes tend to be genetically blessed with a balance in fast- and slow-twitch fibres that perfectly suits their sport. Sprinters have been shown to possess a proportion of about 80 per cent fast-twitch fibres while those who excel in the marathon may have only 20 per cent fast- and 80 per cent slow-twitch fibres.

Other exceptional champions often have unusual genetic advantages: examples include the long legs of most high-jumpers, long-jumpers and pole-vaulters, or the size-17 feet of the Olympic swimming medallist Ian Thorpe. Genes influencing the brain and nervous system are also important. Being left-handed may confer advantages in certain sports. Eight out of ten world motorcycle racing champions are left-handed (usually only one in nine people), giving them a crucial advantage in visual memory and spatiotemporal awareness because of the slightly altered anatomy of a left-hander's brain. Other sports, such as tennis, also have an overrepresentation of lefties. The genes for left-handedness are now being pinpointed on chromosome 2 and may tell us surprising things about the brain. As apes all use the same dominant hand for fine tasks, our human hand variability, and therefore brain flexibility, may have been crucial in brain and language development.

Gender differences in sport are also worth considering. Evolution produced clear differences in physique between the sexes, with men being on average 7–8 per cent larger and stronger overall than women. This increases to about 30 per cent at the hand (as noticed by differences in handshakes). Women were in modern times judged (by men) to be grossly inferior to males in most sports. This was due to prejudice rather than physiology. Only in the last thirty years have women been allowed to run long distances of over

10 kilometres. Previously, it was thought they would suffer permanent injury or infertility. Women are now known to be excellent long-distance runners, and have caught up with men in a big way. The world record for the marathon for men is currently 125 minutes and for women 138 minutes – (close to the men's record in the 1960s) and closing all the time. Undoubtedly our female ancestors were selected like males for walking and running long distances, but unlike men they had to carry children, too. The genetic potential of women in endurance sports is likely to be very similar to males and the gap should continue to shorten.

In general, the genes influencing athletic prowess are likely to involve the heart and lungs as well as the muscles. The hearts of athletes have to be able to work well over a very large range of physical stresses and have the capacity to improve performance in response to training. Genes influencing injury are also crucial. Many sportsmen's and in particular footballers' careers are shortened by tendon and knee injuries. This is likely to be related to subtle differences in genes for posture and elasticity and durability of tendons and muscles.

Finding out the key genes influencing athletic prowess would be very useful to any country hoping to improve its international sporting status. The ACE (angiotensin converting enzyme) genes, involved in heart function and blood pressure are likely candidates. There is a particular variant (the I form) that is present more often in elite endurance athletes (runners, rowers and mountaineers). This allows them to exercise and train their muscles to use oxygen more efficiently than normal individuals. The other variant – the deleted, or D, form – is more frequent in short-distance swimmers. The ACTN3 gene controls the speed and strength of fast-twitch muscle contractions and athletes have a higher proportion of a particular variant (577R) than the average person. Female sprinters have more of this genotype than female long-distance runners. Other genes that may be involved include PPAR alpha, which controls the way the heart muscle uses energy in response to exercise. Another may be the UCP3 gene, which affects the energy-giving mitochondria in muscle, and influences the way different people respond to exercise. Many other genetically determined mechanisms are likely to exist, as well as the genes influencing 'the will to win'.

To be a world champion or record holder requires a unique combination of genes and the right environment. In horse racing, breeding improved speeds and times for about seventy years until they started to level out, with only minor differences between horses. Suddenly, one horse came along that knocked seconds off all the previous winning times and won every race in sight, including the Triple Crown. The horse was called Secretariat and was quickly put out to a lucrative life in a stud farm. He produced four hundred foals – but only one was successful (Risen Star) and none were able to match his prowess. This example shows the crucial importance of the *unique combination* of genes making an individual, rather than merely the number of 'good' genes transmitted from the parents.

There are many similar examples in competitive sport in humans. There are few world champions who win exactly the same discipline as their parents or brothers or sisters. In contrast, there are a reasonable number of identical twins where both of the pair achieve success at the highest level – examples include the Mayer twins in downhill skiing, the Waugh twins in cricket and the Hills twins in horse racing. These identical twins share the unique combination of genes that can make a champion. Recently, two US Olympic gold-medallist runners (Tim Montgomery and Marion Jones) announced they were having a baby – leading to media speculation that they would produce the fastest baby ever seen. While the child will undoubtedly be well above average, the odds are still against his or her winning a future gold medal for running.

Although the right physique and genetic makeup may be essential, some other special ingredient or life event may be needed to produce a world champion. A fascinating study of a pair of identical twins who both competed for their country at the 20-kilometre walk revealed some insights into the importance of personality. One of the twins won the Olympic gold medal while his brother ran only respectable – but never world-beating times. They had lived most of their life together and shared the same trainer for sixteen years. Physically and in physiological tests there were no differences between them. The only slight differences were in personality testing, and in particular responses to and expression of anger. Some life event must have triggered the right anger response that, in combination with the right genes, wins gold medals.

Although none of the genes so far identified are useful enough to predict abilities in an individual, this may soon change. One potential downside to finding these genes is the possibility of their misuse in the future – via illicit and undetectable gene enhancement. A large number of sportsmen and -women in the last forty years have abused their bodies with various drug overdoses, and some have died. Most competitive athletes surveyed confidentially say they would do anything to become a champion – even if the side effects meant they had a high chance of dying within five years of their glory.

My son the bully

M was a lively and physically strong fourteen-year-old boy who was bright and attended a good school. But he had a volatile temper and had often come home from school with wounds from fights. One day his parents received a call from the school summoning them urgently. They were told how their son had kicked and punched another boy so seriously he had to be taken to hospital, still unconscious. They were also informed M had been involved with a gang that had bullied other kids. He was suspended from the school. The parents initially denied there was a problem; then, after some anger, started to accept the truth. His mother felt it was all her fault for not showing him enough affection and emotion. His father also had a bad temper but believed his stricter upbringing and a spell in the army had solved the problem. He felt he was also partly to blame for not disciplining his son enough.

Violent behaviour is as old as our earliest human ancestors and probably goes back even further. Evidence of tribal violence and gang warfare has been seen in our closest relatives, the chimpanzees. Among chimps, bullying, rank and hierarchy are vital to their social units. Yet the propensity to violence and antisocial behaviour shows even larger variation between modern humans, especially males. Twin studies have shown these differences have a genetic influence. A large study of more than 1,500 twins aged seven to sixteen from the UK and Sweden found a clear genetic influence on aggressive antisocial behaviour (such as violence). We asked our adult twins in a different survey directly whether they had ever bullied other children. One in six said yes – producing a heritability for bullying of around 55 per cent.

A predisposition to violence and aggression comes from a number of sources. Levels of testosterone are particularly high in teenage boys and are controlled by a number of genes. There is also an important gene controlling brain dopamine and serotonin levels (via the enzyme monoamine oxidase) that lies on the X (sex) chromosome, explaining why it primarily influences males. The frequency of the gene is highly variable in different populations, which suggests that in certain groups or environments it was favoured in the past. Some of this may have been due to the threat of competition. Groups who lived in sparsely populated areas (such as the !Kung of the Kalahari) had less aggressive instincts and tribal customs than those with more local competition for resources. Obviously, aggression was in many cases a useful trait to our male ancestors, aiding hunting, defeating rivals and impressing females. Nowadays, without a life totally devoted to warfare or competitive sport, these traits are likely to have become less beneficial for everyone to possess – but still, unfortunately, useful for some.

Faced with a child with a predisposition to violence, having a stable home and school life and encouraging peaceful outlets for aggression will reduce the risk. Many bullies are themselves bullied and family and school-based interventions are often helpful. While bullying and aggressive behaviour in adolescents is a mixture of genes and environment, it is not inevitable. Some societies have for a short while produced virtually violent-free societies. Examples include early seventeenth-century New England and early twentieth-century Iceland. It may sound strange, but there is some evidence that aspects of children's appearance or behaviour that makes them victims has a genetic basis. Our twin study showed that the major determinant of whether you were bullied at school were your genes – with a heritability of over 75 per cent. Perhaps in the past, by being submissive and taking a bit of minor physical punishment, these 'victims' managed to avoid serious injuries and death when confronted with more powerful enemies.

Wallflower genes

K was fourteen and although she was well developed physically she had a major problem with shyness. Her parents had noticed that when she was a toddler she reacted badly to new situations and environments. When faced with strangers she was always silent and would never look them in the eye. When boys talked to her she blushed immediately and became embarrassed. She did have a few close friends, but started to lose them when they began to want to socialise more. She was an only child and her parents felt guilty about her upbringing, believing her behaviour was their fault. K herself thought that she was to blame, as she was often told she wasn't trying hard enough to make friends and be nice to people. As she became more upset and alienated from her friends, her parents took her for counselling.

Shyness is one of many personality variations that distinguish us as individuals. It usually manifests itself as altered behaviour such as lack of eye contact, physical signs such as a racing pulse and blushing and feelings of awkwardness or being stared at. When young people are surveyed, 40–50 per cent claim they are shy, and it may be becoming more frequent. Tests in adopted children and twins have shown that around 50 per cent of shyness (or introversion) is due to genes, around 25 per cent to family upbringing and the rest to individual experiences. Studies of shy kids as toddlers aged eighteen months found that virtually none of them became extroverts as teenagers, and two out of three remained shy and fearful as adolescents.

Being shy is an opposite characteristic to risk-taking behaviour and has been found to involve chemical pathways in the brain. Shyness in Europeans has been found in one study to be slightly more common in people who are fair with blue eyes, thin, narrow faces, hot foreheads and faster heartbeats. A single part of the brain that deals with emotion, the amygdala, may be responsible for many of these features under the influence of chemical transmitters such as dopamine and noradrenaline. Shy children may be so because their genes make their brains susceptible to overreact to new stimuli. Culture then dominates and in most environments converts the feeling to a fear that they have done wrong, and so they experience guilt and a lack of confidence that lead to avoidance of these stressful situations in future.

Guilt appears to be a very human emotion, which may have evolved as a social buffer to temper the basic instincts of survival and reproduction.The 'Nordic shy types' may have gained a survival advantage from their genes, being more resistant to cold than darker-skinned more extrovert migrants, with shyness as a by-product. Another possibility is that shyness had some advantages in itself: perhaps by taking fewer risks and not trusting strangers immediately, shy people stayed alive longer in certain environments or scenarios. Males might have seen shyness in females as a future sign of fidelity and therefore found it attractive. In terms of treatment it has been shown that shy patients respond well to therapy when they are told that their shyness is an innate characteristic that they were born with. When they accept this, they find that their self-esteem and confidence improves and they can deal with the problem more easily. People who manifest too much guilt will express more complex problems – as Woody Allen can testify.

All of us have some aspects of our behaviour that we would admit is a bit odd – whether being too brash, shy, stubborn, obsessive, cautious, risk-taking, volatile, anxious or any number of other quirks. Psychologists classify our personalities into five main domains: neuroticism, extroversion, openness, agreeableness and conscientiousness. However you define the traits, they are all clearly genetically influenced – with heritabilities of 40–50 per cent. What is still unclear is how important shared family environment is to personality. Studies of twins have shown that identical twins who end up with rather different personalities in adulthood experienced considerably greater stress as children than twins who remain very similar. It may be that our personalities are moulded very much by the individual ways we react to our environments and families. Personality traits are also related to brain levels of serotonin and dopamine – and we saw earlier how some of these traits could be useful. It is likely that, for our ancestors, having a few neurotic or anxious people in the tribe was advantageous in avoiding danger. Having a few obsessionals probably also helped keep the camps tidy.

5

Genes, Attraction and Sex

Love, sex and conception are the essence of how we mix and propagate our genes and are therefore a pivotal part of life and evolution. How we subconsciously select or attract mates and our sexual instincts are an essential part of our genetic heritage. Although modern contraception and society have recently altered our conscious attitudes to sex and relationships, the genes that programmed our built-in desires are millions of years old and have not had time to adapt to our new environment. Understanding our subconscious desires and drives can, however, help us all deal with the trials and tribulations of a modern love life. The average man in a lifetime produces more than a thousand billion sperm – enough to fertilise most women on the planet many times over. Women on average produce only four hundred eggs in their lifetime. These differences between the sexes are the key to our hidden desires and behaviour.

Fatal attraction

H was 27 years old and had had a number of boyfriends – but none had lasted more than six months. One day she announced to her parents and friends she'd really fallen in love with her current boyfriend. They were very surprised at her choice. She couldn't explain why, either, because in the past she normally went for men with strong character, who were physical and extrovert. He was thirty years old, slightly built and not very sporty, but was quiet, kind and generous

*and had a good sense of humour that made her laugh. She
said he made her feel comfortable and secure. They settled
down in a long-term relationship and had children within two
years.*

Women, like men, have urges and desires to have sex.
Everyone knows that (despite the modern use of contra-
ceptives) the biological purpose of sex is reproduction and
keeping the genetic line going. But why should all of us have
the same strong urge to have sex, which as we now realise,
often indirectly leads to procreation? It's because all of our
ancestors had the same urges and passed them to us. Nuns and
monks (if well behaved) couldn't have been our ancestors and,
however brave, strong, intelligent or good-looking you or your
genes were, without sex it counted for nothing in evolution.

Females of most animal species bear the main cost – in
terms of time and energy – in conceiving and rearing children.
Human females expend an extra thirty thousand calories
during pregnancy, an equivalent of running 650 miles. It is
also a very risky business, as an estimated 1,500 women die
every day giving birth round the world. Most of the deaths are
in developing countries such as those in Africa, where the
risks for a woman are a high one in sixteen – compared with
a still notable one in 3,700 in the US. Human males may
invest only a few minutes of energy in lovemaking, whereas
unsupported women will have, for a minimum of three years
(and usually many more), a dependent child to look after.
Females invest much more in the mating game – they are
therefore more selective in their choice. This makes them the
ones that decide whom to mate with, and they have certain
in-built subconscious strategies to help them satisfy their need
for desirable qualities they seek.

The first subconscious female strategy is selecting males on
the basis of their 'good genes'. This is a preference for men
based on indicators of health such as size, strength, facial and
body symmetry, broad shoulders and slim waist, athleticism,
domination and determination. These are genes that

originally favoured their children's survival against disease and adversity and were probably originally useful in the male role of hunting. The concept of symmetry is particularly interesting. It is a signal that, while the body was developing, it was relatively free of parasites and disease. Faces give many clues to symmetry, but subconsciously women are also attracted to men with symmetrical hands, feet and other body parts. Symmetrical males are more successful and have on average greater numbers of sex partners and lose their virginity earlier. Recently it was shown that 'symmetrical' males had more concentrated, mobile and healthy sperm than asymmetrical males. In a related study, female ratings of male attractiveness correlated strongly with the quality of their sperm, suggesting a clear biological link that women may have evolved to tune in on. The average male has shoulders that are 20 per cent wider than their waist. Males with higher shoulder-to-waist ratios have been found to be more attractive to women and they mature earlier – a clear signal to women.

Women may also subconsciously select men as being 'good providers'. This is based on signals that they will make good fathers – such as kindness, reliability, generosity, status and wealth. These measures mean they are more likely to hang around after fertilising the woman and provide food and protection for the mother and children. Worldwide surveys of all cultures from Alaska to the Amazon to Australia have shown that women desire an ideal mate to be on average three years older, taller and of higher status than themselves. These are all characteristics of the good provider and appear to be an in-built preference rather than based on purely cultural influences.

For the female, there is often a difficult balancing of desires between the two types of male, which depends on her circumstances and culture. H, in this instance, having had a number of liaisons with 'good-gene' males, fell for the good provider as a better choice of long-term mate. Indeed, males have responded in evolution to female wishes to demonstrate traits that females will recognise as signs that they are likely to be committed and will look after both them and their children. These include signs of status such as a prestigious job, a flash car or a smart suit and being generous.

One problem with this cunning plan is that males are aware of it. 'Good-gene' males with attractive symmetrical broad-shouldered physiques may often try faking the 'good-provider' characteristics such as boasting about their status, exaggerating their generosity and commitment, and conning females to get their way. If they succeed, they are more likely to spend their time competing with and fighting other males and abandoning their mates early. They've also been shown to have more extramarital affairs. Females, in turn, have become, via evolution, more astute in spotting the fake provider – and it's easy to see how an escalating arms race ensues.

Hormones also play an important role in females' choice of a mate. Studies have shown that menstruating females will prefer symmetrical angular male faces (masculine good gene type) only at the most fertile (mid) point of their cycle, whereas during the rest of the cycle they will generally prefer more feminised, rounded and softer male faces (good providers). Other research shows that first impressions can be crucial. It seems the female brain remembers and rates the first image of the male in the original hormonal context. Even if she meets him a second time at a different point in the cycle, her preference on his suitability as a good provider or someone with good genes will still partly depend on her cycle at the time of the first meeting – thus many subsequent bunches of flowers may be wasted but both types of males are in with a chance.

Do genes smell attractive?

While she was on holiday in the Med, V quite liked M when they first met at a bar. Her pale skin and blonde hair contrasted with his dark skin, handsome features and dark curly hair. He had a good symmetrical face and athletic physique – but she was somewhat put off by his over-enthusiastic use of cheap aftershave lotion and resisted his initial advances. Two days later when they met again by chance in a café at lunchtime, he was wearing smart clothes but no aftershave. After a few minutes of idle chat and eye contact, she suddenly felt very physically and emotionally attracted to him. They ended up making love later that evening and for the rest of the week.

Women have an amazing sense of smell, most of it sub-conscious. They consistently score higher than men in tests of sensitivity and discrimination of smells. Using innate genetic mechanisms (ADAM family genes), women select potential mates by their subliminal smells – a skill that varies at different times of the menstrual cycle. These aromas are linked to the detection of certain immune recognition genes called the HLA (Human Leucocyte Antigen) family on chromosome 6. These HLA genes are carried on the surface of all our cells as unique identifiers, like barcodes, and are important in recognising foreign intruders. In carefully controlled studies women tend to prefer men's smells on the basis of HLA genes that are different from those inherited from their fathers. If the genes match too closely, the smells make them feel comfortable but are not 'sexy'. If the genes don't match at any level, women also don't fancy the male. Women on the pill behave as if they were pregnant and will tend to prefer smells that are more familiar and less' exotic' or sexy.

The power of genes for scent and attraction is shown by a recent experiment for a BBC documentary which involved two identical female twins who were asked to independently smell the natural scents of six anonymous men and rate their preferences. They then met the same six men for five minutes each as fast 'dates'. Amazingly, both women independently preferred the same scent and what they thought was the same man. However, the scent they preferred belonged to two men – identical twins. Both women had, unknown to them, been smelling and rating three pairs of identical male twins – each pair being genetic clones.

Although not in the same league of sensitivity as women, men can also pick out the HLA genes of potential mates fairly well and avoid those that are too close and potentially incestuous. Experiments in stickleback insects have shown that, as well as avoiding inbreeding, the more diverse the HLA mixture, the better the chances that offspring will survive parasite infections. Geneticists have recently identified at least five hundred different genes for smell in humans and another five hundred that have now ceased working (pseudo-genes). This means there is incredible diversity between any two humans in our sensitivity and response to smell. These human sniffing skills pale into insignificance compared with those of other animals, particularly rodents.These hidden talents in us all have evolved from times when recognising your kin was essential for protection, reproduction and avoiding inbreeding – and ideal mates were those whose genes were a bit foreign (but not totally alien). Perhaps this in-built desire in women to mix genes around a bit explains the allure of the mysterious stranger to women in most cultures.

Many men and women wear perfumes and scents that have been used for at least five thousand years and have changed relatively little. Eau de Cologne has been around for two hundred years and Chanel No. 5 since 1921. But are these scents disguising our true smells and merely fooling the opposite sex? Experimental studies of perfume users have shown that the scents we prefer to wear and we'd like to smell on others are related to our own HLA immune genes – the relationship being very strong with powerful scents such as musk. So we probably do subconsciously select fragrances that match our individual genes and augment our natural smells. This is one extra reason why buying perfume for someone else can be a risky business.

What do men want?

F, a 35-year-old divorced salesman, was having a quiet drink in a hotel bar after a difficult day on the road. He was working hard at the moment and went out on episodic dates, some arranged via the Internet, but didn't have a steady girlfriend. He considered himself fairly good-looking for his age, and despite being a bit flabby round the waist, kept himself in reasonable shape. The bar was fairly empty and he engaged in some light conversation with the waitress, who was in her early thirties, friendly but a bit plain and chubby for his tastes. After a few more drinks he found himself alone with her in the bar at closing time. The way she 'accidentally' touched him suggested she was keen on something more than just conversation. He was very aroused by her flirtatious behaviour and her generous breasts, and started to find her quite attractive. After the bar closed he invited her to his room and they made love twice, before she left him early in the morning. They did not meet again.

Several experiments have been carried out on US campuses involving an attractive actress who plays a stranger who in a public place flatters men and asks them for no-commitment sex – in these situations 75 per cent of men agree. This contrasts with women, who hardly ever agreed to sex with a male actor in the same situation. This may be that women are more used to the situation – or perhaps women are just more perceptive than men to being 'set up' – as no student (as far as we know) actually got to have sex with the teaser.

Whatever the reasons, men rarely turn down a good sexual opportunity (and, in evolutionary terms, one that allows them to spread their genes) and have correspondingly low thresholds and standards in their choice of short-term mate. In multicultural surveys of male preferences, youth and attractiveness are universally desirable. Female signalling of sexual availability and promiscuity is a major plus factor for casual sex – but a big turn-off for long-term relationships. The only major turn-offs that particularly affect men are women who demand excessive commitment, lack sex drive or are too hairy. Studies in singles bars have found that men also find women increasingly attractive the later in the night they meet them, as their chance of sexual success for that day declines – even accounting for the effects of alcohol. Lust is certainly driven by men's subconscious drive to inseminate. Studies have shown that men decrease their ratings of their female partner's attractiveness as soon as ten minutes after sex.

Although females ultimately end up doing most of the mate selecting, male behaviour and initiation strategies depend on their own inherited impulses. Men universally desire females who show the greatest indirect signs of fertility and ability to bear them many children. Men therefore generally prefer women who are youthful, healthy and attractive to other men. Studies of men's desires in sixty different countries have shown very consistent patterns. Across all continents the average man will both desire and marry a female who is three years younger than he is (matching female preferences). As he ages, and his social status improves, he will desire a relatively younger and fertile mate. Women's fertility peaks at around eighteen and is fairly negligible after the age of forty.

To pick a healthy mate (one likely to be free of actual or future disease), males look for signs such as plentiful shiny hair, smooth disease-free skin and an energetic personality. Men also look for signs of health and fertility in a woman's build, with an ideal waist–hip ratio of 70 per cent. Men may be put off women with large waists, as this simulates pregnancy – and commitment to an already pregnant female would be a genetic disaster. Men, much more than women, value 'attractiveness' in a mate. Part of the reason is a sign of health, the other reason is as a status symbol. Men (particularly if unattractive) who obtain sought-after partners are perceived as being more successful by their peers. The same is not true of women. This increased prestige, combined with the fact that women can still find physically unattractive men with power and status sexy, perhaps explains the stranger liaisons of 'ugly' male politicians.

The one-night stand

While H was at college, she found herself having a few drinks at the bar with an older, rather arrogant but still attractive lecturer in his late thirties who was showing off and entertaining several other female students with his wit and humour. He managed to impress H with his passion for poetry and songwriting and was buying drinks generously. Later that evening he invited her to come back to his place and then into his bed. Lovemaking was over quickly with plenty of passion but without much foreplay – and without her reaching orgasm. He was satisfied and promptly went to sleep. She returned to her room soon afterwards, disappointed, frustrated and feeling guilty about her behaviour. She couldn't understand how she'd wanted sex with this man, who (she now thought) wasn't really her type.

As previously discussed, for men the genetic (subconscious) drives to have casual sex are fairly easy to understand since they have little to lose and much to gain genetically in terms of potentially spreading their seed widely. It's more difficult to understand what drives women to have brief sexual liaisons. In women the desires underlying the criteria for selecting a short-term mate are similar to those for a long-term mate – but the emphasis shifts more to the signals showing good genes, and standards (particularly of commitment as 'good providers') are allowed to slip a bit.

Other subconscious reasons why women desire short-term sex range from simply wanting the obvious pleasure of sexual gratification to seeking to be fertilised by 'good genes'. These genes would produce a child with the same good and 'attractive ' genes, who, like his or her father, will continue to have success in the mating game and thereby propagate the woman's genes. This is known as the 'sexy son' theory. Other reasons why women have casual sex are related to the need to practise and experiment with mating skills – improving flirting and sexual techniques and testing the quality of males they can attract. The menstrual cycle and hormonal balance may also play a part, as women are more likely to look for a casual sexual partner during the few days of maximal fertility before ovulation. Another theory for casual sex comes from observations of behaviour of apes and females in primitive populations, and it holds that women try to secure additional males to protect them in times of trouble or as a potential extra source of food and resources.

The underlying sexual desires are predominantly genetic, due to evolutionary pressures, but these desires vary in intensity between individuals and even within women at different times of life. Our twin study found a genetic component to one-night stands. Genetic desires are constrained by different cultures and circumstances such as age, availability of males and what the competition are doing. Women, as in the above scenario, may also sexually select men on the basis of their creative skills (art, music, words and humour) because this suggests (rightly or wrongly) that their brains and therefore their genes are healthy. This is likely to be one reason why humans evolved 'non-useful' or – for practical survival purposes – superfluous creative skills in the first place: because women found them sexy. Women nowadays – chasing male artists or comedians solely for their supposedly healthy genes – may well be disappointed.

Sexually incompatible

D moved in with B, but after the wild sexual frenzy of the first few weeks, when both of them had amazing energy and desire, the frequency of sex diminished steadily. She thought his interest in her body had declined. She had always had a high libido and sexual appetite. She (unbeknown to him) had had around twenty previous partners, while he told her he had had three. She couldn't understand why he didn't want to have sex at least every night. He said he didn't want to be pressured – but denied any real problem. She found this disturbing and thought he had found someone else. They broke up a few months later and eventually ended up with different long-term partners.

Sex is universal but sexual libido, promiscuity and masturbation vary throughout life and are highly variable in both women and men even within the same cultures and environments. Twin studies from our group have shown that the frequency of both actual and desired sexual intercourse and number of sexual partners has a considerable genetic influence with heritabilities of 30–60 per cent.

Incompatibilities are to be expected – and may not necessarily be related to a breakdown in the emotional relationship. In general, males say they want more sexual partners than females, the average male college student desiring (in theory) about six partners over the next year compared with an average of one for girls. Over a lifetime the ideal figures were eighteen partners for males and five for females. Some of these differences will be cultural – but a recent study of 16,000 subjects from 50 different countries found similar and consistent results, with an average of 6 partners for males and 2 for females over the last ten years. Expectations of frequency of intercourse also vary. After four years of marriage about 40 per cent of men complain of a lack of sex compared with 18 per cent of women. Males in all countries report that they fantasise about sex twice as often as women, and fantasies more often involve multiple anonymous participants.

In recent large-scale confidential surveys the average (mean) number of actual lifetime sexual partners reported in the UK in the 1980s was ten for males and three for females – similar to surveys from the reportedly oversexed French (eleven and three partners respectively). In the most recent UK

2000 survey, it increased slightly to thirteen and seven respectively. However you interpret these numbers, men appear to have more lifetime sexual partners than women. If this is so, where do the missing females come from? One obvious explanation is male boasting. The other explanation is female 'selective memory'. Studies have shown that while males respond to sexual questions similarly in different situations, females vary in the frankness of their responses depending on the confidentiality of the survey. In an anonymous situation without fear of disclosure, and subject to a mock lie detector, females admitted to more sexual activity – approaching male levels. There is an unwritten rule that to obtain the true numbers of sexual partners, the reported number needs to be divided by three in men and multiplied by three in women. The surveys suggest that a magic factor of one and a half would do the trick for most people without resorting to the use of a lie detector.

The problems with averaged sex statistics are the odd ones out – the tiny fraction of people with the out-of-control libidos. A better figure to report is the median, which gives the commonest number and discounts the greedy people with hundreds of partners who distort the figures. When you calculate medians, the rates are more similar for the genders – being six partners for males and four for females. In population surveys, men come out as the more promiscuous with around one in a hundred men admitting to more than a hundred partners. Don Juan is reputed to have had 2,065 sexual conquests, easily matched by modern-day sports stars such as the basketball player Magic Johnson, who confessed to thousands (and got the AIDS virus) or a multitude of male rock stars.

Female record holders in the sex stakes are more publicity shy than their male counterparts, but, in fact, easily outperform men. Prostitutes boasting more than a thousand paying men a year are quite common, as are porn stars such as Annabel Chong, who held the world record for the number male sex partners in 24 hours (251). Her efforts came to little, as she was beaten two months later by an even more energetic Ms St Clair, who reached three hundred. Interestingly, the male volunteers were happy to work unpaid – unlike the women. Famous historical females such as Cleopatra are said to have had many men, but the Marquis de Sade's heroine, Madame de St Ange, may be the record holder with ten thousand to twelve thousand lovers in twelve years. Prostitution is not that uncommon, with nearly 1 per cent of US women having had sex for money at some time; and, in 2000, 4 per cent of UK men admitted to having paid for sex in the past five years. It is estimated that around a third of men in urban areas will visit a prostitute at some time in their lives, and the US Kinsey report found that 15 per cent of men regularly used prostitutes. Some of the gender differences in numbers of sex partners may be due to this secret black market.

Hormonal differences may also play a role in sexual desire. Male sex drives are driven to a large extent by the male hormone testosterone, which, although it has a daily pattern, being higher in the mornings, is fairly constant throughout the month. Women are influenced by a mixture of oestrogen, progesterone and testosterone in a complex manner and are more likely to want sex close to the time of maximum fertility, at mid-cycle. Differences in hormone levels may explain important age effects – with males' sex drives diminishing after eighteen and females' peaking around thirty, and, as the hormonal differences between the sexes diminish, sex drives become more similar after the age of sixty. One of the largest recent national surveys of eleven thousand people from the UK showed that the average male *and* female (the two figures actually agreed) between sixteen and forty-five with a partner have sex around six times per month. Oddly, here, there was no major decline due to age.

Despite the age-related differences in desire, in nearly all human relationships the frequency of sex drops off rapidly, usually halving after the first year and declining more slowly thereafter. Similar analogies exist in other animal species. A good example is the so-called rooster or Coolidge effect, named after a joke between the thirtieth president of the USA, Calvin Coolidge, and his wife after they had separately toured a chicken farm. On seeing the rooster vigorously mating and being told he did it dozens of times per day, the first lady asked that the president be informed of this fact. On later hearing this, the president asked, 'Always with the same hen?' to which the guide replied, 'Oh, no – we change the hen each time.' 'Please tell that to Mrs Coolidge,' said the President. The male rooster can mate more than thirty times a day. However, he can do it with the same female only five times. He gets increasingly bored and loses interest. But put a different female with him regularly and he'll carry on all night. The male animal is instinctively programmed to not waste sperm on females after he is sure to have inseminated them. It's unlikely that any interesting experiments like this have been done in humans, but for women lack of novelty in sex may not be so much a problem for them as it is for men, who like to disguise their females in lingerie and other exotic costumes.

The reason men in general want (or say they want) a greater number of partners than women may be related to their theoretically unlimited reproductive and genetic potential. The highest official number of children sired by a man is 888, by the emperor of Morocco, Moulay Ismael the Bloodthirsty. Unofficially, several of the Chinese emperors are likely to have had more partners, as they had harems of 2,000 young women who were on a rota system to coincide with their maximum fertility. In reality, modern paternity tests might have shown that many royal helpers (the non-eunuchs) might have assisted when the boss got tired. Women, in contrast, rarely manage more than five live children, although the world record is held by an energetic and tough Russian, Mrs Vassilev, who (if we believe it) had 27 successful pregnancies and, with many twins and triplets, a total of 69 babies.

Within these extreme male and female stereotypes lie large variations and differences, based both on culture and our genes. It may be hard to distinguish what we say we want from what we *really* want. Studies of primitive tribes where women have freedom and financial independence show that the females are as naturally promiscuous as males. When it comes to sex, there's no such thing as Mr or Ms Average. Sexual compatibility is a fairly crucial part of most couples' relationships – as problems account for about a third of marriage breakdowns. Unfortunately, there's currently no easy way of gauging a long-term prospective partner without a bit of risky trial and error.

Orgasm and the G spot

H was in her early twenties and in a steady relationship with her slightly older partner. They had sex around three times per week. She climaxed very occasionally and only when she stimulated herself. She had had one previous partner and was not very sexually experienced and obtained most of her sexual information from magazines such as Cosmopolitan. *Eventually, she broached the subject with her best friend, who, in contrast, claimed to climax every time and sometimes had multiple orgasms the same night. H felt pretty sure her partner may be letting her down sexually. But not having many men to compare her partner with and not being able to ask his previous girlfriends, H was also worried that it might be her own inadequacies.*

Whether women suffer like men from sexual dysfunction is an area of considerable current controversy. Proponents say that it is a major disease area kept secret by male taboos and indifference. Critics argue that normal human variation is being fabricated into a disease by pharmaceutical companies to sell Viagra-type drugs to women. As usual, the truth probably lies somewhere in the middle. There is enormous individual variation in female orgasms compared with male ones, both in the type and frequency and the amount of stimulation needed. At one extreme, in recent large UK population surveys around 30 per cent of women said they had difficulty achieving orgasm and 12 per cent have never experienced one. At the other extreme are rare cases of women who are reported to be capable of having up to a hundred orgasms in one hour.

No one would argue that key elements in female orgasms are variable emotional factors, related to cultural and environmental influences – the right time, mood, place etc. Nor would it be right to diminish the influence of the male partner, whose attitude, staying power and technique are clearly important. However, there is still a factor missing to explain the one women in four who finds achieving orgasm so difficult. To our surprise we found that in our twin survey a considerable proportion of the large variation in orgasmic ability in females is due to genes. The similarities among identical twins for ease of orgasm or for difficulty reaching orgasm were greater than for fraternal twins. The results were the same whether we looked at regular sex or masturbation, with heritabilities of 40–50 per cent.

Another old (male) theory was that female sexual pleasure and orgasms were nonbiologically relevant optional extras – as no other animals had them. This is now known to be false. Orgasms have now been recorded in a wide variety of female animals including the African sooty mangabey monkey, who stimulates herself with her hands during sex, the female orang-utan, who uses tree branches as sex toys, and the infamous female chimpanzee raised by humans who *apparently* masturbated to a naked male *Playgirl* centrefold.

Current theory is that the higher threshold for orgasm evolved in some women so females would be more discriminatory about finding a long-term mate – one who would have to try harder to keep her happy. His sexual abilities, patience and staying power would also be reliable indicators of his health and prospective genes, as well as his likelihood of hanging around and being a good dad. On average (if such a thing exists), a woman needs fivefold longer stimulation times to reach an orgasm than a man – an average of around thirteen minutes versus two and a half minutes.

These ratios give men plenty of opportunity to show off their skills and patience.

An orgasm may also have developed as a biological mechanism to help the body. Research is sparse in this area but it may reduce the chances of infection by clearing away debris and bacteria from the cervix and vagina. Women who have regular orgasms may also have improved fertility rates. The mechanism may be that the orgasm allows more sperm to be retained, altering the cervix and helping sperm swim up through the womb and into the tubes more effectively. To obtain the full fertility benefits the woman has to orgasm at the earliest sixty seconds before, or within forty-five minutes of, the man's ejaculation – not always an easy thing to time to perfection.

Differences in women's G spots could be another reason for variation. The G spot (a term coined by Ernest Gräfenberg in 1950) refers to an area a few centimetres up inside the vagina on the side closer to a woman's stomach. Buried in the flesh here are the Skene's glands (the female equivalent of the male prostate gland). In women, these glands are also thought to produce a watery substance that may explain female 'ejaculation'. The tissue surrounding these glands, which includes the part of the clitoris that reaches up inside the front wall of the vagina, swells with blood during sexual arousal. There's also some evidence that nerves in the area produce an orgasm different from one produced by clitoral stimulation.

Studies are finding closer similarities between male and female sexual mechanisms. The clitoris has been found to contain the same types of molecules (nitric oxide) that stimulate erections in males. To control the process is a key enzyme (called PDE5), which is blocked by the drug Viagra, and the drug can assist orgasm in some women. Evidence of high levels of PDE5 in a part of the body would suggest that area has a special sexual-stimulation function. Although some doctors are still sceptical about the existence of the G spot, studies recently found the enzyme PDE5 in the vagina of five volunteers. Another study, which dissected the vaginas of fourteen dead women, revealed that the key PDE5 enzyme was mostly clustered in the G spot. But in two of the subjects (14 per cent) with much lower concentrations of PDE5, no Skene's glands were detected. For such women, having a vaginal orgasm may be anatomically tricky, if not, impossible.

We can't overlook the role of the brain in pleasure. Studies of excited females have shown certain parts of their brain light up on sexual stimulation or orgasm. A number of areas are involved but the hypothalamus seems central to the process. As with other pleasurable sensations, the reward chemical dopamine is a major player. How dopamine is released and exerts its effects is likely to be influenced by a number of different genes. Finding out what they do opens up clear possibilities of manipulation of orgasm by drugs that affect brain chemicals.

Whether the genetic influences on orgasm act primarily on psychological or anatomical factors is uncertain. But, given the wide variation between women, some may need to get their men to try a lot harder – and to prove themselves genetically worthwhile.

Male infidelity

H was a 32-year-old woman who was still happily married after four years with two happy and healthy children. One day, quite out of the blue, she found out that her husband J, who was two years older, was having a steamy affair with a younger work colleague. When confronted, he said the relationship was purely physical and was just a fling. They reconciled and she forgave him. Two years later she found some sexy text messages on his phone from a new mistress. He eventually confessed to having had a number of affairs. She threw him out – and they divorced acrimoniously. She was later reminded by her mother-in-law that J's father had been similarly inclined and had several affairs. They wondered whether J's father's influence had been important.

Infidelity is a genetic trait older than history. The skeletons of a ritual murder of an adulterous couple and cuckolded husband were found recently from twenty thousand years ago, demonstrating that the infidelity strategy was often fatally risky. Even if they survived death or injury, they may have faced the major threat of social isolation. This may have weeded out those who overdid this strategy. Its occurrence varies in different cultures, but most surveys show that 20–50 per cent of males have been unfaithful at some time. The UK 2000 survey showed that 15 per cent of men admitted having been unfaithful in the previous five years. In our male twins, 22 per cent reported having been unfaithful at some time. One of the first links of sexual infidelity to 'sin' was made by the ancient Hebrews: 'Thou shalt not commit adultery.' However, it was apparently quite OK for men to bed prostitutes, widows, concubines and maids – only married women were off limits.

The amount of promiscuity in males is difficult to gauge in averages. The 1 per cent of men who are the most active account for 16 per cent of female sexual encounters – meaning that a few men are very successful at juggling partners and the rest are not. Historically, it has obvious survival advantages for your genes. Our male ancestors of a hundred thousand years ago are likely to have fathered children from a number of different females. However, given that at that time there was only one human per square mile, the chances of bumping into a young fertile available female were pretty slim. Some males would have hung around and helped provide for the mother and kids, while others left as soon as the woman was pregnant – to try to find pastures new. Staying faithful and hanging around was probably also not without risks and genetic pitfalls. The woman may have been infertile due to infections, may have been unfaithful, pregnant with another man's baby, or had a 50 per cent chance of losing the baby in childbirth or infancy. The faithful dad would have also had to wait a while until he could reproduce again. Before baby foods, mothers breast-fed much longer and were usually not fertile again for three years.

Historically, not all men (and their genes), therefore, could afford to be too patient, and, if the opportunity of a young fertile female presented itself, they would take it. A breeding strategy based on quantity of partners rather than quality was at face value a good one – as long as some of their children survived. The genetic line of men who were too single-mindedly faithful, who hung around even though they couldn't reproduce, soon died out. Yet there were also downsides to the totally unfaithful strategy – the survival rate of children would be much lower and many may have been killed by new male lovers; unfaithful women would also run the risk of being injured or ostracised in a rather small community.

A survey of modern women showed that 44 per cent aged under thirty said they would 'drop' their husband if he had an affair, a proportion that dropped to 28 per cent of women in their forties, and 11 per cent in their sixties, either showing how male fidelity is more important to young women when they are at their most fertile, or maybe reflecting their greater economic independence and choice.

It's clear that, for modern man and his ancestors, having the makeup and impulses to encourage them to be unfaithful, albeit to different degrees, was a genetic advantage. This all seems rather unfair on women until you realise that men need someone to be unfaithful with – most women are willing participants and most of them are married too. So what's in it for them?

Female flings

V was thirty and considered herself relatively attractive. She had been reasonably happily married for eight years and had two kids. Her husband was 35 and initially financially well off, due to family money. But his work had not been going well recently and they had started to argue. Their sex life was regular and predictable, and they usually had sex once a week at weekends. One day she was alone in their large house with the 21-year-old window cleaner, a law student on his holiday job. V found him physically attractive and she found herself seducing him. She led him upstairs and after the first few minutes of his first frenzied attempt they had sex in a more satisfactory and leisurely fashion and on four more occasions. She never saw him afterwards, but two months later found she was pregnant. Her new baby looked slightly different from her other kids and was brighter. Her husband never suspected he might not be the father – and V never knew for sure.

Five to forty per cent of women in most Western cultures admit infidelity. Previous surveys have ranged widely in size and quality, and depend on the population and how they were selected. A recent survey by our group of five thousand UK women aged eighteen to sixty showed infidelity rates to be around 23 per cent – similar to the earlier Kinsey report in the US. Real rates have been estimated to be 10–30 per cent higher than those reported, due partly to guilt and selective recall of those surveyed. The lowest infidelity rates are in strictly controlled groups such as orthodox Jews or Muslims. Fantasising about infidelity is, however, nearly universal in both men and women. Our twin study of five thousand UK females has confirmed infidelity to be also a genetic trait – explaining about 55 per cent of the differences between women.

Worldwide surveys have shown that between one in ten and one in fifty births within a marriage is due to female infidelity (also called nonpaternity). It is more common in women who marry young men of low wealth and less common in women of the same age with more elderly rich husbands. Recent surveys suggest that in well-off societies with contraception, rates are probably less than 2 per cent. From the evolutionary perspective of the woman, her tactic is to obtain the best-quality genes she can from a mate – provided she can ensure their survival. Thus a good female genetic strategy is to mate with a man of better genes (more intelligent, creative, handsome or healthy) and allow the existing partner to bring them up and protect the children.

Some primitive cultures in South America embrace a system whereby women have different fathers of their children (multiple paternity). They believe, as our ancestors perhaps did, in (incorrect) stories that babies are made by a mixture of sperm from different fathers that end up eventually as a whole person. Tests in some human tribes and a wide variety of animals have shown that, when females choose multiple fathers, the children survive better than when all have the same father. This suggests that a mix and diversity of genes is somehow beneficial and supports the view that women may have affairs subconsciously to improve the genes of their offspring. A woman could also not rely on one partner as a sole provider and protector for life and she had to hedge her bets, both for her own and for her genes' survival.

Studies have also shown that women tend subconsciously to have more extramarital sex in the middle of their cycle prior to ovulation when they are at their most fertile. They also tend to pick more masculine-looking mates. When being unfaithful they also tend to have greater numbers of better-timed orgasms to coincide with their partners' ejaculations – so increasing chances of conception. This subconscious strategy of timing and orgasms greatly increases chances of fertility in a lover even if they have sex with their regular partner more often. Female infidelity also tends to peak just before their reproductive abilities start to decrease between the ages of 31 and 40, supporting the theory that new male partners may subconsciously be a backup if they want to jump ship and start again. Careful studies of sandpiper birds have shown that the females, like most birds, are very good at deceiving males (and ornithologists) and having discreet affairs. Genetic testing found that females were more likely to cuckold their males when their own DNA was more closely related to that of their steady partners. In other words, they were more likely to seek genetic variety by having an affair –

and let their steady partner bring up the other, 'more foreign', male offspring. This mechanism of seeking out the ideal adultery partner, as with humans, probably involves smell or pheromones (as discussed previously) as well as the eyes.

In Charles Darwin's Victorian era it was believed that women and most female animals were naturally coy and men were invariably the more sexually proactive. With the advent of genetic testing, it has been shown that females from the majority of species are not monogamous, many being very good at having a 'quickie' behind the bushes while their regular mate is catching a worm. Women are usually more successful at managing infidelity covertly than men. They have more to lose from discovery, and, given their superior communication abilities, have better skills at deceiving males who lack female levels of awareness and intuition. They are therefore less likely than men to be discovered. It is now believed that our female ancestors had sexual equality up to relatively recently – before farming and male domination took over around ten thousand years ago. Women before this era often had different lovers and fathers of their children – whom for the most part they chose voluntarily. It is in this environment of groups of thirty to fifty people living together – who may have met other groups every few months for a 'party' – that we need to think of how female sexual practices evolved over the last hundred thousand years.

Good evidence of women's secret fickle nature – and that men in the past had to compete sexually for the favours of women – comes from a close look at male sperm. Sperm have evolved a wide range of defence mechanisms to try to ensure that one of their number beats off the opposition and is successful in fertilising the egg. This is achieved by having a variety of sperm that have various roles – some lasting longer, others swimming faster and large numbers of sperm acting purely as blocking or killer sperm to try to stop other men's semen fertilising the female.

It is believed that even today sperm competition is quite common, with around 4 per cent of births conceived as a result of two different men's sperms fighting it out inside the female. We get to see the results of this clearly in only rare cases of non-identical twins born simultaneously to different fathers of different races, where, for example, one baby is black and the other white. The quantity of male sperm is also a guide to the fidelity of female animals. Gorillas keep harems and their women on a tight rein. Their testicles are small and therefore sperm production is quite low. Chimpanzees are promiscuous and have extremely large testicles and large sperm volumes. Humans lie somewhere in between, suggesting that a small degree of female infidelity has been around for a long time.

The other (nonsperm) contents of seminal fluid may also contain magical 'jealous' properties to stop females mating with other males. The final drops of the human male ejaculation contain a mild spermicidal agent. Some animals' sperm, like that of mice, contain a substance that blocks up the female's vagina for several days with a cement-like plug. Others such as fruit-flies, have sperm that contain anti-aphrodisiacs, and the house fly has semen with poisonous toxins, ensuring that the female can mate only once. In humans a recent study found that women who were having regular unprotected sex were happier and more content than those using condoms – suggesting the possibility that human semen contains unknown substances that promote happiness and wellbeing and reduce depression. These 'mellowing' chemicals may be an evolutionary tactic to keep females content with their current partner and reduce infidelity.

Most women in various surveys disapproved strongly in principle of infidelity by either sex. We found this belief to be strongly culturally influenced and certainly not genetic – even in the one woman in four who had been unfaithful. This looks like a case where culture attempts to keep our natural genetic tendencies in check. So, if genes for infidelity exist, what about gene therapy? Genes controlling the brain hormone vasopressin from faithful prairie voles have been transplanted into unfaithful mice, resulting in faithful mice by influencing the same areas of the brain involved in addiction. So what about humans? Could gene transplants replace the prenuptial agreement?

Jealousy and competitiveness

F married at eighteen and was happy for several years, but his wife became increasingly jealous. This mainly involved his female work colleagues, with whom he occasionally went out for a drink or a meal after work. He explained that he didn't find them attractive and was merely being sociable and liked them as friends. His wife started various rumours about the promiscuity of his female colleagues and tried to belittle them and mention their unattractiveness whenever possible. She started spying on him and waiting for him outside his office. She said she wanted him to change his job. He denied ever having an affair, but she continued to provoke arguments. The strain became too much and their marriage broke up soon afterwards.

Jealousy is another powerful characteristic that is part of being human. It has been designed by evolution to counterbalance and control infidelity. Despite going hand in hand with love, it has today a strong stigma attached to it. Studies we performed with female twins showed that, given a scenario of male infidelity, nearly all women were jealous to some degree, but the extent was under genetic control. Men and women are equally jealous, but about different events and have different thresholds. Men get very jealous over the thought of sexual infidelity and being cuckolded (a man's worst-case scenario: his partner's impregnation by another man). For women, so-called 'emotional' infidelity, such as the thought of losing or reducing the care, protection and resources of the partner to another woman and child, in most countries, is more important in inciting jealousy and rage.

Both men and women suffer extremes of jealousy and this is the commonest cause of domestic violence, accounting for over 10 per cent of murders in Western countries. Jealous violence is greatest in young men who have young attractive female partners but relatively little wealth or power themselves. At least 80 per cent of domestic murders in Europe and the US are triggered by male jealousy. Older well-off men who perhaps feel more secure are less likely to succumb to violent jealous rage. Under English law, if a husband or wife kills the other, a mitigating plea of provocation can be used, which often reduces the sentence from murder to a few years for a lesser punishment. Similar laws exist in other countries. As jealousy underlies most domestic murders, the use of the plea is under review in a number of countries – as it appears to condone jealous violence. The law dates back to medieval times, when it was intended to allow discretion for men duelling over women in a more ritualised way of settling disputes.

Jealousy exists because, as we have seen, men and women unfortunately can't trust each other due to infidelity. A key part of this lack of trust is the secret and variable menstrual cycle of women. Ovulation is very variable. It is commonly believed that a woman is most fertile on Day 14 and pregnancy can occur only in the five days including and preceding ovulation. However, recent studies of women have shown that only one in three women will be fertile during Days 10–17 of her cycle, and 1–6 per cent of women can get pregnant on Day 1 of their period – commonly believed to be the safest. A small percentage of women even ovulate twice in a cycle. This variability is such that males (and females themselves) can't consciously detect optimum fertility. Other apes and mammals have clear cycles and signals (such as pink bottoms) to tell each other when females are both fertile and available. Outside these times they are not interested in sex and are unavailable and therefore 'safe' from other males. In contrast, women are always sexually available. In humans, to be sure of fidelity, you would have to guard females all the time, which is impossible in nonoppressive societies, as females are always worth approaching by men.

The art of keeping your mate away from rivals takes many forms, ranging from violence and threats, to being so good to them (including providing regular sex) that they couldn't possibly leave. It can also involve separating them from social situations involving other potentially dangerous rivals. Recent studies have suggested that, subconsciously, males pay more attention to their mates and guard them more jealously just before ovulation. It's not clear yet how they pick up these signals, but, as males prefer the scent of T-shirts worn by women who are ovulating, it may be due to smell. In addition, men have another subconscious strategy: using jealous sperm. If fertile women have been away from the partner for over a week, and potentially in contact with other males, the male will rate her as more attractive, have sex with her and, after deeper thrusts, produce double the usual number of sperm (an already generous 350 million). This is a bid to neutralise any competitor's sperm that may be inside her with his blocking sperm and his killer sperm. If he's seen her all day and is sure of her fidelity he will be meaner with his donation. A total lack of jealousy may not be healthy, either. Women in particular take this as a sign of a lack of interest and commitment and it is often a cause of marriage breakdown. Given that it's such a complex process, it is not surprising that few of us are able to find exactly the right balance.

Cads, dads, sex and food

M was twenty, single and attractive in a quiet way. She had many good-looking male admirers in the small town where she lived who tried to court her. Their approach was usually to take her out for a meal (usually a cheap Italian restaurant) and give her lots of wine. After a goodnight kiss and an attempted grope in the car on the way home, they would usually ask if they could come upstairs for coffee. She said no, not until she got to know them better. Many didn't call her again; some tried once more – then usually gave up, saying they'd had better offers elsewhere. M despaired of finding a man she could trust who wasn't in a hurry. Finally, one came along – a rather shy man she met through work. They went out for four months before she eventually took the initiative and asked him if he wanted to stay the night.

There are many different strategies to sexual and reproductive success. The balance of the sexes makes a difference to female choices. When there is a shortage of males in the population, females get less fussy and when there is excess, females are pickier. There are four stereotypes in the animal kingdom, which have been used in real-life and computer simulations of evolution. These are cads (philandering males), dads (faithful males) and 'coy' and 'fast' women, the proportions of which will also make a difference to each person's and each sex's subconscious strategies. In these stereotypes, cads are impatient and abandon females as soon as they've mated; dads will patiently woo females for long periods of time without sex in order to prove their future faithfulness; coy women mate only with males they think will be faithful; and fast women will mate without any courtship or proof of fidelity. Our scenario involved a coy female battling against cads until she met a dad.

As soon as any two stereotype groups are the majority, the alternative genetic strategies become more successful. For example, if the majority is a monogamous society where the two groups are coy females and dads, a few mutant genes of fast females will be very quickly successful and replicate rapidly, as they won't waste valuable time courting. When the fast women eventually take over, they will in turn be duped by the mutant cads, who take over from the dads. The fast women would find themselves holding the baby on their own, with less chance of survival, and die out. Computer simulations have worked out that if three-eighths of males are cads and a sixth of women are fast, the system is in temporary equilibrium – which is close to the estimates of cads and fast women in modern society. In reality, it is likely to be a constantly changing balance between the groups – one increasing until the others take over. The computer studies also showed that the same situation held true even if men were for example cads for only three-eighths of the time or women fast for a sixth of the time.

Few women will admit to contemplating having sex within a short time frame of meeting a man. Surveys of college students in the US in the early 1990s showed that women invariably prefer to wait longer than men to have sex. After being given a range of different theoretical scenarios from five years to one hour, at six months women gave a 50–50 likelihood of having sex and after only a week said it was highly unlikely and less than an hour virtually impossible. Men, on the other hand, said that after a week they were more likely than not to consent and less than an hour was quite possible. The UK 2000 survey showed data of what actually happens in practice nowadays, with 43 per cent of women reporting having had sex within a month of meeting their last sexual partner.

For the first 4.5 million years of man's existence, before the discovery of fire, food was raw and cold and it's hard to know what particular morsels would have most impressed females. There are many examples of male animals routinely exchanging food for sex. Females are generally happy to play this role. Male hunters of the Aché tribe of Paraguay commonly bribe potential mistresses with meat. The male scorpion fly gives the female a delicious wrapped insect to eat. If she accepts it, they both hold on to it while she eats it and he mounts her. When she finishes the meal, sex is over. It's hard to see this catching on in Italian restaurants, however.

Male problems: all in the jeans?

J suffered from premature ejaculation. When, aged eighteen he started to date girls, he would regularly ejaculate in his underpants on starting to indulge in petting, even before his girlfriend touched him. Eventually, with time and experience, he slowly improved and lasted longer – and was able to penetrate and inseminate women relatively normally. However due to his oversensitivity, having sex was always a stressful experience for him and his girlfriends. Although his penis was of a normal size, he felt it was on the small side, which added to his insecurity.

For the essential components of male reproduction, penile dimensions and quantity of sperm, there is a narrow range that evolution has given men to work within. If penises are too large or oddly shaped they won't work properly and won't easily enter the female vagina. If penises are too sensitive or not sensitive enough, they won't deliver the semen needed to produce children. Those males with penises that 'failed' their owners died out and are not our recent ancestors. Our ape ancestors didn't hang around. The average chimp ejaculates after seven seconds, bonobos fifteen seconds and gorillas a full minute, making the average four minutes for a man quite luxurious and lethargic.

In a recent UK survey a third of men (34 per cent) reported having a current sexual problem of some description and the majority of men have suffered performance problems at some time. The most common problems reported are erectile dysfunction and premature ejaculation. Premature ejaculation (PME) is very common, affecting 11 per cent of men. While anxiety and stress obviously play a major role, and relieving these help in treatment, innate or genetic factors are also responsible. It may seem contradictory that a sexual problem has a genetic basis. But, in evolutionary (not romantic) terms, depositing one's seed quickly (if unsatisfactorily) was often a safer tactic than hours of dangerous lovemaking, where you could be attacked from behind, replaced by a more aggressive male – or, worse, your female mate got bored and abandoned you.

Impotence or erectile dysfunction (ED) is also very common, affecting at some time one man in five, but for obvious reasons permanent impotence is not likely to be genetic, as it can't be passed on. Contributory factors such as anxiety, depression, vascular disease and smoking all have genetic components. Occasional or late-onset impotence and premature ejaculation, as reported by males in our twin study, do have a genetic component with a greater similarity in identical rather than nonidentical twins.

Our ancestors used to have small bones in their penises (as most monkeys still do) – and didn't have to worry about sustaining an erection. Why men evolved to lose this bone is less than clear. One theory is that prehistoric women didn't want all men to be the same and wanted to judge their health or good genes by their ability to maintain an erection. Keeping an erection without a bone requires much more effort – and means that the nerves and blood supply and vascular pumps have to be in good working order and the male in good shape.

The various vessels, pumps, hormones and chemicals needed to get and maintain an erection are complex, and different chemical pathways can have the same result. The process starts off in the hypothalamus section of the brain, which releases chemicals to open the blood supply that fills the organ, while valves close off the exit of blood, keeping it where it's needed. All this is regulated by a number of chemicals. Any interference in this delicate mechanism means the system and the organ quickly collapse. One such crucial chemical is nitrous oxide, now famous because the drug Viagra, by chance, was found to block its actions (via the enzyme PDE5) and became a billion-dollar-a-year earner for the company Pfizer. The search for the key genes influencing erection and ejaculation and the chemicals that alter other pathways is likely to be a rewarding business.

Why males of our species have such large, flexible penises is unclear. Despite our insecurities, human males have the longest and thickest penises of any primate, which in a recent US condom survey was reported to be an average of 5.78 inches (14.7 cm) long, and just over 5 inches (13 cm) in previous European surveys. In the animal kingdom size is roughly related to promiscuity. But human male penises are smaller than those of many other animals, such as the wild boar (18 inches, 45.7 cm) and the bull elephant (5 ft, 1.5 m) or the promiscuous Argentinian lake duck with a corkscrew, like 12-inch (30-cm) penis. Contrary to popular myth, gorillas have penises only 0.8 inch (2 cm) long (and pencil thin) that nonetheless seem to do the job quite adequately – perhaps because male gorillas are very muscular and use their strength rather than their penises to keep harems of females happy. In animals in general, the physical size of males is inversely related to the size of the testicles, as the physically well-built hunks use their physical size to keep females in line rather than fill them with sperm.

Male testicles have evolved to produce large quantities of sperm to help ward off competition, but penis shape probably evolved out of female preferences, i.e. those that stimulated them best. Females chose mates with certain-shaped penises and abandoned the rest to history. Another interesting and unusual feature of human penises in evolutionary terms is the width and shape of the tip, or glans. The coronal ridge beneath it is designed (while in the thrusting mode) to remove any sperm from other males that is already in the woman's vagina. Experiments have shown it does this very effectively if the thrusts are deep and fast. Other animals have even stranger-shaped penises to do the job, including attachments such as spikes, shovels and spines. So women shouldn't complain – it could be worse!

Love genes

When H first met S, she attracted his gaze by looking at him, then lowering her eyes, flicking her hair and moving her head to the side in the Princess Di manner. Eventually their eyes met, their pupils dilated, their faces flushed, their stomachs and hearts fluttered and their lips and sex organs filled with blood. Lust was in the air. They had sex – lots of it – spent several months continually obsessed with the other's every move and thought, only seeing the good points of each other. They then decided to live together, became an 'official' couple, stopped seeing their friends for a while and did everything as a pair. Six months later S conceived. They married and had two more kids over the next six years. They supported each other through her serious illness and later when he briefly lost his job and became badly depressed. They are still happily married thirty years later.

Although this reads like a fairy story in the modern world, couples like this are still quite common in most cultures. What is the biological basis for these strong bonds between unrelated individuals? Flirting patterns are seen in all animals, and in apes the behaviours are quite similar to our own. In most cases it is the female that initiates the process, with the male picking up the clues, but believing (usually falsely) that it is he who is taking the initiative.

Once the flirting process has brought together the couple, and set up the crucial eye contact, chemicals come into play. One common scientific interpretation of the process of being in love breaks it down into the separate emotions of lust, infatuation and attachment. All of these probably involve separate chemical signals acting on different parts of the brain, which are under genetic control. The three processes are all selfishly designed to ensure successful reproduction.

The lust stage of the original attraction is related to getting you ready for action via short-term visual, smell and hormonal signals, involving pheromones, adrenaline and testosterone and leading to dilation of the pupils, facial flushing and transfer of blood to other parts.

The infatuation stage is the bonding stage with a partner, which means you think of nothing much else for a minimum period of three months up to one year. The drive for this comes from a number of other hormones and chemicals, including natural amphetamines (PEA), dopamine and serotonin. With this natural 'high', it's easy to understand how lovers can stay awake all night, talking and caressing, a state likened to 'permanent anaesthesia'.

Since the average time for the healthy young couple having regular sex to conceive a child is three months, one can easily see how lust and infatuation have evolved to last a similar time. The real test comes later, if and when the attachment stage kicks in. Here, the infatuation wears off and the true nature of the relationship takes shape. The attachment stage classically lasts three years. It is designed to protect both any offspring and the mother, who with a young child is most vulnerable for the first three years. Oxytocin is one of the brain chemicals likely to be responsible, and is found in higher levels in couples in the first few years. This chemical is also raised during lovemaking and important in mother–baby bonding during breast-feeding, giving a feeling of security and wellbeing. Endorphins are also likely to be involved. These are natural opium-like chemicals, which help dampen anxiety and provide a sense of security. Vasopressin, a hormone linked to pair bonding and fidelity, is another.

While falling in love seems to be a fairly universal phenomenon, some do it more often than others. Known as 'attraction junkies', these poor individuals crave the attraction phase and get attracted to unsuitable partners, are briefly elated and then get a major 'down' afterwards, before seeking another partner. These people have an inbuilt (genetic) need for their brain to be constantly stimulated by these 'feel-good' chemicals, either because they have lower levels in their bodies or their receptors are less sensitive. There are major similarities with other addictions. Drugs that increase these chemicals in the brain seem to help them lead more normal lives. People are attracted to each other in different ways – some are drawn by physical lust and others ideals of romantic love. Some researchers looked for genes for romantic love in twins. To their surprise they found that it was mainly influenced by culture and upbringing. This suggests that, while we are genetically programmed to fall in love, how we do it will depend a lot on our circumstances and upbringing.

Monogamous relationships have obviously evolved in parallel with promiscuity, and have their own long-term rewards. Individuals, particularly men in successful long-term relationships who have been able to fight off temptations and deal with married life when the chemicals run out, are much less likely to die of depression, to commit suicide or to suffer major psychiatric disease and tend to live longer. Couples also give each other mutual support in achieving higher social status and financial rewards by sharing the workload and having a greater ability to network.

Is a successful marriage genetic?

P was a successful doctor who at the age of 28 married a nurse named C, whom he met and fell in love with at the hospital where they worked the year before. They got on well for the first two years – all their friends agreed they made a perfect couple with everything going for them. After then, things started to go wrong, with increasing rows over petty and trivial matters. Eventually they stopped talking together completely and he moved out of their flat. They officially separated and a year later acrimoniously divorced. They never saw each other again. Within two years both had found new partners and they settled down, remarried and both had kids and lived happily ever after.

Tolstoy succinctly described a successful marriage in *Anna Karenina*: 'Happy families are all alike, unhappy families are all unalike in very different ways.' There are many reasons why a marriage breaks down that may be nothing to do with sexual attraction. Any problem with finances, in-laws, jealousy, infidelity, ill health, change of status or unemployment, religion or children's upbringing can be enough to break up a relationship. One would expect, therefore, that divorce would be a completely environmentally or culturally induced event. Surprisingly, divorce has been shown to be partly (50 per cent) genetically determined in both men and women based on a number of twin studies in Europe and the US.

However, divorce has a major cultural component. It is now a universal phenomenon, affecting all societies and groups. This has been true only in recent times. In the last ten thousand years since agriculture and farming started, men and women got married to each other and had to work as teams for their children to survive. This meant that if a couple split up, they risked losing the land and their children. For most of recorded human history (but only a fraction of evolutionary history) marriage has been a way for men to own women – discarding them if they were infertile or unfaithful. These hard facts made divorce pretty impossible except for a few elite and wealthy couples. However, in very recent evolutionary times (the last hundred years), patterns have started to change. Divorce is fast becoming the norm in Western society, with rates approaching 50 per cent in most countries. The UK has the second highest rate of divorce in Europe and as a consequence around one in ten fathers has no contact with his children. Social observations in many countries show that divorce becomes more common when the environment of women starts to change, such as when they acquire financial power and independence. This may reflect the unmasking of innate genetic tendencies.

An insight into what married life was like before farming comes from looking at the primitive hunter-gatherer tribes with more female sexual equality. The !Kung of the Kalahari have divorce rates of 40 per cent and in Paraguayan tribes the average couple will change partners eleven times. This suggests divorce is not a recent phenomenon, and may have been commonplace in our distant ancestors. Most men like the idea of polygamy: 84 per cent of world cultures allow men to take more than one wife at a time, although in these societies only 5–10 per cent can actually afford to. In polygamous societies most women end up with a mate – although they may have to share him. Usually, when given the choice, women in these cultures opt to be a second wife of a successful man (with more guaranteed resources than the only wife of a poor and unsuccessful one). For males, the reality is that apart from a lucky few, most end up worse off, with less chance of ever having a mate.

Why so many modern marriages break down is obviously due to a complex mixture of events, culture, gender differences and personalities. Surveys have shown that the top two reasons in all cultures are infidelity and infertility. We have seen how the tendency for either partner to stray into infidelity is under genetic influence, constrained by cultural or religious factors, and this obviously leads to marriage breakdown in some circumstances. Infertility is the other major factor. It is in neither partner's interest genetically for them to stay together if conception fails. Overall, only around one-twentieth of mammals stay together as couples. Many bird couples used to be regarded as happy monogamous relationships. However, ring doves are now known to have an annual 25 per cent 'divorce rate', which usually occurs when the couple fail to produce offspring in any one breeding season. A United Nations study of 45 societies found that across countries 39 per cent of divorces occurred in those couples without children, compared with 3 per cent when there are four or more kids.

A twin study examining reasons for divorce found per-sonality traits to be crucial. Divorce was greatest in those with the most extreme emotional personalities and differences in these traits explained around a third of the genetic influence on divorce. Other genetic factors that might have a role are differences in paternal interests in child rearing, communication skills, sexual compatibility and jealousy or total lack of it.

The one common factor to divorce in all countries and cultures is the timing of when couples break up – the clear peak is three to four years after marriage. This coincides with the time that the attachment chemicals wear off and when a healthy child would start to be independent and out of danger. They should have named the famous Marilyn Monroe film, *The Four-Year Itch*. All animals weigh up the potential benefits and costs of leaving a partner and the risk of starting again. For women with children, the risks are greater than for men, as with increased age and children they are 'less genetically attractive'. Slightly fewer women than men manage to remarry, although what is remarkable is that, despite all the heartache and trauma, the vast majority of us (over 75 per cent) fall in love again and remarry an average of only three years later. Humans are suckers for serial monogamy.

Gay genes

Mrs P was delighted when her only son (her husband had died early) was successful at his boys-only religious school. He was never very interested in sports and was quiet and non-competitive. Nevertheless, he started to date girls and on a few occasions had sex with them – although this was not a totally satisfactory experience. When he was at university, however, Mrs P later found out her son had sexual relations with men and he later 'came out' and admitted having a regular boyfriend. Being from a conservative and religious background, his mother found this difficult to accept. This led to friction and strong feelings of guilt about her role and his environment and upbringing in his sexuality. She had read that homosexuality was due to an overbearing mother and an underinvolved father and felt perhaps that there was some truth in this.

Homosexual experiences in men are very common. The Kinsey report of the 1940s in US men claimed that as many as one in three men had had some form of homosexual encounter. But this now appears to have been a biased over-estimate. A 1970 US survey reported that one in five males had reached orgasm with another man, but only 7 per cent after age nineteen. The national UK 2000 survey reported that 5.4 per cent of men had had a homosexual relationship. The number of men who have regular male sexual partners is lower, however, at around 2–3 per cent. Few men are exclusively homosexual. At least 50 per cent of those reporting homosexual experiences have been married at some time, and the majority report bisexual experiences. Homosexual experiences are commoner in early life in those with higher educational attainment. Homosexual and more frequently bisexual behaviour is common across a wide spectrum of the animal kingdom – making it seem unusually rare in humans. Primates such as the highly promiscuous pygmy chimpanzees (bonobos) mount each other regularly, baboons fondle each other and some male sheep are interested only in other males. Male dolphins, who have similar-sized brains to ours, are known to copulate with each other as well as with unfortunate and reluctant passing male turtles.

Theories of the 'causes' of homosexuality have always been controversial. In the 1960s it was hypothesised that over-bearing mothers, distant fathers and too many older brothers were the cause. In the 1980s the role of exposure in the womb to sex hormones (particularly testosterone) was also hailed as the cause. However, studies failed to show any clear hor-monal differences in adult homosexuals and heterosexuals – and giving homosexuals male hormones just made them more promiscuous. It is possible there is a critical time of hormonal imbalance in the foetus – but this simple theory seems unlikely given that if the trait was caused in the womb, all twins would have the same sexuality, which is not the case. One recent study shows homosexual and bisexual males to have longer index fingers than fourth fingers, which, although unconfirmed, suggests some developmental effects may have occurred in the womb.

Over the last fifteen years a number of twin studies and studies of brothers have shown fairly consistent results that male homosexual tendencies are around 50 per cent heritable. Each study was announced to a fanfare of publicity and controversy. The chances that an identical twin would be gay if his brother was were around 50–50. The chances that one brother would be predominantly homosexual if his brother was gay were around three or four times higher than expected, around 10 per cent. The effect of family upbringing probably has only a very minor role, if any. One pair of male identical twins in our study both had heterosexual relationships with women when they first went to university – but one found the experience was lacking something, and subsequently experimented with men and found he preferred them. The other twin has remained heterosexual. Neither could explain the reasons for the differences between them.

Several studies have tried to find causal genes. In 1993 one group claimed to have found a gene linked to the X (sex) chromosome based on only forty pairs of brothers, but the findings couldn't be replicated in a larger follow-up study in Canada. Contrary to the theories expounded in the early nineties, there is unlikely to be a simple 'gay gene' – but predisposition is likely to be due to a combination of many genes of moderate influence interacting in a highly complex manner. Studies to find the genes for sexual orientation attracted too much publicity and in the US no further funds were given to pursue the work.

Are there physical or biological differences in homosexuals and bisexuals to make them act differently? Animal experiments – as in exclusively gay rams – have shown differences in the size of an area of the brain controlling emotion: the preoptic nucleus of the hypothalamus where male hormones are produced. These studies confirmed previous studies of males who died of AIDS who had smaller control centres in the hypothalamus than heterosexuals. Male homosexuals show less 'male-style aggression' and on average, in tests of motor-performance skills such as throwing a ball, score intermediately between females and heterosexuals. Certain chemical signalling substances – pheromones – may also be involved. Studies have shown that these chemicals in fruit flies can turn insects in either direction – a form of future gay aftershave perhaps.

If we look at our history, we see that male homosexuality has been commonly recorded as being socially acceptable, and not just a recent phenomenon. The ancient Greeks and later the Japanese samurai held homosexuality in high esteem, both artistically and militarily – believing that male soldiers fought better if they were in love with one another. Different cultures since, as we know, have held different views on its place in society. Obviously, for males who never reproduced, their genes could never be passed on, and so we have to talk about the role of genes that influence bisexuality and what led them to survive selection and evolution.

These genes must have had some indirect survival or reproductive advantages in our ancestors. Benefits for repro-duction might have included being more attractive to women, either physically or creatively, or by exuding more female qualities of empathy and communication. Sexually, they may have been more skilful in foreplay or more precocious in chatting up and communicating with women. The survival advantages may have come from their ability to bond with other males or by their superior communication skills which may have helped them to avoid violent arguments with other males.

Another theory is that bisexual genes confer increased fertility when passed to females, so increasing the number of children with these genes, or that the genes are linked to some other hidden survival trait. These advantages must have ensured that their genes continued from one generation to the next, despite the disadvantages of extra risk of disease and occasional risks of violence and social ostracism. Being in a minority group may also have been important as an aid to success. Gay men interestingly have the same general (non-gender-specific) taste in partners as heterosexual men, i.e. younger, fit and good-looking. They are also in general more naturally promiscuous. Before AIDS one survey of the San Francisco Bay area found that 75 per cent of gay men had had more than ten partners and 25 per cent had more than a thousand. The prolific activities of gay men are perhaps because they are not constrained by female partners' genetic and cultural instinct for commitment. In other words, hetero-sexual men would do the same if they could and women behaved the same way. Unfortunately for straight men, most women don't behave like the stars of male pornography.

One rather unusual New Guinea tribe, the Sambia, actively encourages homosexuality in all males for several years from puberty till marriage. Once the men marry, however, any further gay activity is frowned upon. This tradition seemed to have no side effects on the tribe and perhaps strengthened bonds of loyalty as well as improving sexual skills. Apparently, 10 per cent of New Guinea tribes and some Australian groups had similar customs. These examples in unusual societies and the commoner forms of adolescent experimentation such as are rumoured to occur in British public schools suggest that – given the right culture and environment – most men can be flexible in their orientation.

AC/DC

C was a petite thirty-year-old divorcee with two kids. She was sporty and had a normal upbringing with two older brothers, whom she enjoyed playing with. She never had many sexual partners but considered she had, overall, had a reasonable sex life. She recently met a rather extrovert female student, E, aged 21, who offered to help her look after the kids during the holidays. One day over coffee, E confessed to finding C very attractive. Although initially shocked by the first kiss, eventually C succumbed to her friend's seduction, and found

*to her surprise she enjoyed the lovemaking. They stayed
together for three years, until C started secretly dating other
men again. She found she also still enjoyed relationships with
men in a way she couldn't explain. C had told her mother
early on in the relationship with E, and she was supportive –
but felt guilty that she hadn't dissuaded C from playing boys'
games and hadn't encouraged her to play more with Barbie.*

Female homosexuality is not as common as the male
equivalent, and has attracted less scientific interest and
controversy. Recent estimates from large surveys in the UK
report around 8 per cent of women with lesbian experiences
and 5 per cent with a sexual partner. As for males, there are
numerous examples of lesbian behaviour in animals,
including the infamous pygmy chimpanzees, as well as
macaques, who give each other orgasms and fight over
females with males. Then there are gulls that form lesbian
couples to bring up baby gulls together. Lesbian-type
behaviour may help females to be more fertile when
impregnated by males. Like men, over 80 per cent are
bisexual rather than exclusively homosexual, and also, as
with men, it is primarily a genetic trait. Studies in sisters and
twins in the US, Australia and the UK, including our twin
study, have shown a clear genetic influence ranging from
40–50 per cent, with little influence of society or upbringing.
The chance that a sister of a lesbian will have the same
bisexual tendency is around threefold or 25 per cent, but only
about 5 per cent will be exclusively homosexual.

Unlike male homosexuality, lesbian behaviour tends to be
evident later in life with 50 per cent of experiences reported
after age 25 and 25 per cent after 30. Also unlike the case of
men, there seems to be a gradual transition of preferences
from males to females, with high numbers of women with
only mild to moderate lesbian feelings, with fewer all-or-
nothing individuals. There is no clear evidence of their being
more aggressive than totally heterosexual females – although
in one study they had better ball-throwing skills.

A sign of potential for homosexuality in men and women is so-called 'gender nonconformity' as a child (boys playing like girls and vice versa), which is also mainly a genetic trait rather a learned one. The advantages in evolutionary (genetic) terms of female bisexuals are broadly similar to those of males. These characteristics include probably being more skilful sexually; they may be better at dealing with heterosexual relationships and are well equipped to have secret successful affairs without disrupting the couple. In our past history, by having lesbian affairs women may have strengthened bonds with the powerful women in the tribe and given them greater protection and resources, greater status and therefore access to higher-status males.

Intriguingly, among lesbians, normal female hormonal drives are still important and women at mid-fertile point of the menstrual cycle are more likely to initiate sex and reach orgasm with other females than at other times. Overall, bisexual females tend to have children earlier than, but in similar number to, heterosexuals, explaining to some degree why the genes have survived. When asked about their ideal partner, most lesbians prefer other women who are slightly older with status, suggesting they still have 'normal' female preferences, with only the gender being different. They are in general no more promiscuous than heterosexual women – with most having fewer than ten steady partners, suggesting they have the same natural instincts for commitment as heterosexual women.

As for men, there is evidence from animals about the flexible nature of homosexuality. Teleost fish and the marine goby can all change sexual preference and characteristics in different social settings. These changes are triggered by hormone changes that alter the structure of the brain in the preoptic area of the hypothalamus – the same area altered in human homosexuals.

A man's best friend?

J loved his dog C. Ever since they met when C was a two-week-old puppy and J brought him home four weeks later, they'd had a special relationship. They went everywhere together, on holiday, walking, fishing and golfing, and the dog slept at the end of his bed. When his new girlfriend moved in, all three of them were very happy. After a few months, however, she started to resent the time he would spend alone with C – talking to him, playing for hours and even sharing his meals with him. He didn't seem to have enough time for her, but always had time for his dog. They started having rows about it, but he refused to change his ways. She left shortly afterwards.

Why do dogs have such a special rapport with humans? Is it instinctive or just their training? We now know from recent genetic and archaeological research that dogs evolved from domesticated wolves some fifteen thousand years ago in China. Most of the huge global variety of dog breeds that we see today – ranging from the minute Chihuahuas to the Great Dane – came from only three tame wolf bitches, domesticated by early Chinese farmers. This means that man domesticated dogs long before even cattle, sheep and goats. He probably used them initially in hunting, and then selectively bred them for qualities such as tameness. The less tame ones probably ran away to join the wolves. As time went on, dogs have been bred increasingly for their social skills and their calm nature to enable them to interact with humans. The selection process may be picking animals whose genes are frozen in childhood mode, making them more uninhibited and playful. This process has been amazingly successful over just a few hundred generations.

Recent tests have shown that dogs perform better than our closest relatives, chimpanzees on tests of interaction with humans and reading subtle human cues about the location of food. This is despite having smaller brains. A good indicator of intelligence is the brain-to-body-size ratio. This ratio of brain (in grams) to body weight (in kilograms) is highest at 7:1 in humans, 3:1 in chimps, 1:1 in cats and 2:1 in dogs. What is more remarkable is that these superior communication skills are also found in six-week-old puppies never exposed to humans – so they must be genetic, not taught. There may be nothing special about the wolf. Breeders in just a few generations have managed to produce tame foxes that are starting to look remarkably like dogs.

As we discussed before, for behavioural traits, the genes set the scene, producing the recipes for a few basic instincts and key survival strategies. Once these are in place, the brain via its billions of neurons and connections and guided by these genes, is especially receptive to learning in response to particular stimuli in key areas. These stimuli can also in turn increase or decrease the influence of individual genes at crucial times of life. These animal-breeding studies show the amazing influence the selection of genes can have on behaviour and social skills over a few hundred generations. Men and women have been selecting each other for breeding for hundreds of thousands of years, and, as we can see, our system is still far from perfect. A genetically modified wolf can still quite easily get between a man and a woman and pursue its own strategy.

6

Grown-up Genes, Instincts and Risks

Having found your ideal genetic mate or realised that you made a big mistake and you prefer to be on your own, you have more time to worry about other problems that modern society throws at you. The way you react to these new challenges depend on genes from your ancestors, who didn't have the same temptations or environment. The genes you inherited were often meant to protect you from completely different situations and scenarios.

Can't give up?

F was 35 and successful and had smoked cigarettes since the age of sixteen. He had tried to give up but failed several times, but now he was determined. His wife, E, had managed to break the habit relatively easily and he knew it wasn't good for his health, as his dad had died from lung cancer. After a week without nicotine, he became very agitated and short-tempered, and had rows with his wife. He succumbed and restarted smoking. He felt inferior and weak, blaming his lack of willpower, but didn't want to show it. Since giving up cigarettes, E found she was drinking about six or more cups of coffee per day and found she needed this quite badly in the mornings. For a bet with her husband she tried to give up coffee for a week, and was

141

surprised how hard it was – she had headaches and felt tired all the time.

An estimated 1.2 billion people regularly smoke cigarettes and one in two smokers dies a premature death due to heart disease, cancer or one of many other illnesses. Men have a one-in-four risk of dying of lung cancer. Lung cancer in women is now overtaking breast cancer as a cause of death – thanks to women's recent susceptibility to tobacco advertising and the fact that in the UK twice as many girls as boys under eighteen smoke. One in four adults in Western countries now smokes. This figure has declined gradually from the 1950s, when 56 per cent and 70 per cent of men in the US and UK respectively smoked, often encouraged by the habits of their doctors.

Seventy percent of smokers now say they want to give up, 46 per cent try each year and 38 per cent have sought advice. Some people need maximal help such as nicotine patches, support groups or hypnotism, while others can give up effortlessly. Starting to smoke, usually as a teenager, is due to a combination of environmental and social factors as well as genes. However, continuing to smoke in the long term and difficulty in quitting is under major genetic control. Twin studies have shown that genes explain around 70 per cent of the differences between people. This explains why some quit easily and others struggle. Racial differences related to these genes have also been found. The average Chinese smoker uses fewer cigarettes and inhales less nictotine per cigarette than his or her Western counterpart – perhaps explaining the lower lung-cancer rates. A direct link has been found between the amount of craving for cigarettes and levels of nicotine in the blood. For addicts, the lower the level falls, the greater the craving. Genes probably control an individual's nicotine levels.

A number of genes have now been associated with nicotine dependence and related to consumption. These include those that control the dopamine reward pathways in the brain that give the pleasurable sensations and cravings, such as the DBH and MAO genes, variations in which lead to a threefold risk of addiction. Different variations of a dopamine transporter gene polymorphism (SLC6A3–9) determine how easily people can give up. This gene was found to influence novelty-seeking behaviour and a need for the brain to have external stimulation. Unfortunately, there is little consistency between studies in different populations. In a recent summary of 46 genetic studies, only the gene affecting serotonin (5HTT LPR) came out as a consistent risk factor for smoking addiction in different populations.

In studies of drugs such as Zyban (which interferes with dopamine) for smoking cessation, participants with a decreased-activity variant of the CYP2B6 gene reported greater increases in cravings for cigarettes following the quit date and were about one and a half times more likely to relapse during the treatment phase. This gene has been found to affect both nicotine metabolism and the effect of the drug, altering levels of dopamine in the brain. Such effects could contribute to withdrawal symptoms and bad moods after quitting, thereby promoting relapse.

The commonest drug we take that affects the brain is caffeine, which is used by 80 per cent of the world's population, in coffee, tea and carbonated drinks. The average American consumes around 225 mg of caffeine daily. Brewed coffee contains around 160 mg, instant coffee, 90 mg, tea 60 mg and fizzy drinks around 40 mg. Caffeine is like any other drug; taking it gives you a temporary high and stops the brain relaxing and falling asleep. It does this partly by blocking the release of relaxation chemicals such as adenosine. People who start using a lot of caffeine need more and more to keep their alertness levels up as their adenosine brain receptors increase in response. Once levels get high, side effects occur and stopping can be difficult, and, as with hard drugs, you get withdrawal symptoms such as headaches, restlessness, anxiety, drowsiness and nausea.

Some people seem to need more tea and coffee than others. Some people can drink unlimited quantities of coffee without affecting their sleep, while others can't sleep after smelling the aroma. Twin studies have confirmed that these differences between people are strongly genetic and not due to family environment or culture. Heavy caffeine drinking (more than 600 mg a day or seven cups of instant coffee) is the most genetic habit with a heritability of over 70 per cent, but with the side effects of drinking too much or withdrawal also being influenced by genes. Some of these genes are likely to be similar to those influencing addiction to cigarettes and alcohol, but others are specific to caffeine – and are as yet unknown.

While coffee or tea has relatively few serious side effects if taken in moderation, the same cannot be said for cocaine and heroin. When a cocaine user inhales the drug, it blocks receptors in the brain, which normally limit the amount of natural dopamine released. This has the effect of temporarily maintaining high levels of the pleasure hormone in the brain. As cocaine use continues, the brain produces more receptors and higher and higher doses are needed to get any pleasure response. Tolerance occurs with all drugs, whether alcohol, cigarettes or harder drugs. Drug users can compensate by increasing their levels a thousandfold. This is trickier for the beer or coffee addict, where bladder size is a restraining factor.

Addictions are nothing new. There is good evidence that our ancestors sought out similar stimulants. Aborigines have been using for thousands of years, a plant called *pituri,* which is rich in nicotine and so helps those who take it to endure desert travel without food. This plant contains three times as much nicotine as cigarettes. Betel nut, a source of cocaine, was being chewed at least thirteen thousand years ago by Polynesians in Timor and coca leaves were chewed more than five thousand years ago in the Andes. These stimulants probably helped our ancestors survive extremely harsh conditions for short periods of time, such as crossing freezing mountain ranges – a far cry from a morning cigarette at the thought of a tough day at the office.

Drunk in charge of your genes

*K was outgoing and sociable but couldn't handle alcohol and
got drunk very fast. She usually developed a bright red face
and chest and felt sick quickly. This caused her many
embarrassing incidents she later regretted – particularly with
regard to the opposite sex. She couldn't understand why this
happened to her, as she usually had less to drink than her
female friends, who seemed to manage without too many
problems. Her dad liked to drink beer regularly but her mum
never touched alcohol.*

Alcohol is broken down by enzymes in the liver, which vary
markedly in effectiveness between people and races. The
immediate side effects of alcohol are due to variations in the
main enzymes, which are known as alcohol dehydrogenase
(ADH). These break the alcohol down into a toxic compound
known as acetaldehyde, which makes you feel sick. There are
five different types of these ADH genes. People with certain
forms of these genes – type 2 and type 3 – convert alcohol
much more rapidly to acetaldehyde. Many people from
American Indian, Japanese and Chinese races – perhaps from
their common Asian roots – have genes that don't produce the
necessary enzymes. A small glass of beer can have a dramatic
effect, causing flushing and drunken behaviour. Some people
feel so nauseous that they never touch alcohol again,
accounting for many teetotallers. Carriers of this gene are
naturally discouraged from consuming much alcohol.

In Japan, virtually all the chronic alcoholics studied have
Western-type alcohol (ADH) genes. Flushing genes, in
contrast, seem to protect their owners completely. Europeans
can also inherit some of these Asian genes, causing wide
individual variation in response to alcohol. Studies in
alcoholics from Finland and Japan have suggested that
variations in a gene controlling the GABA receptor may be
responsible. GABA is a key chemical in the brain. Medical
and governmental advice about average safe levels for health

and driving are useful population guidelines, but often misleading for individuals, as people are different in the way their bodies eliminate alcohol.

Bad reactions to alcohol are often a natural protective mechanism against addiction. Hangovers are caused by a combination of three events: the build-up of the acetaldehyde (which hasn't been effectively converted to acetic acid); a build-up of other toxins from the drink itself; and dehydration. Some annoying people rarely get hangovers and others suffer badly after one glass of red wine – again, this is likely to be genetic, related to the genes controlling elimination. As many of us know, hangovers get worse as we age, due to our weakening enzyme action. Some hangover effects are not due to alcohol itself, but to the other chemicals. There are many by-products of natural yeast that can cause a throbbing head, such as sulphur dioxide added to kill off unwanted bacteria and neurotoxic amines produced by bacteria. Scientists in the field of genetic modification are working on engineered yeast that will be much purer and hopefully cut out some of the nasty side effects.

For poorly understood reasons, teetotallers are more likely to die of heart disease or strokes and have thinner bones than moderate (but not heavy) drinkers, and whether this is due to the benefits of alcohol per se or related to the underlying genes is unknown. Some people believe red wine has some special health properties and hi-tech wine producers are genetically producing white wine with the same oxidising benefits.

If you tolerate it well, you are more at risk of getting permanently addicted to alcohol, which like cigarettes is also strongly genetic. So have a careful look at your family's fondness for the bottle and cigarettes in order to gauge your own risk. If you have a strong family history, this is another good reason not to start – as you won't be able to experiment safely and will find it hard to stop. If it's too late, it is important to admit your genetic weakness and take all the psychological and chemical help you can get to quit. Perhaps in the future, health warnings on the labels will be specific to your addiction genes, breathalyser tests will be adjusted, and individual targeted therapies will become available.

As fast as we identify and find assistance for one addiction, new modern ones will appear to replace them. Good recent examples are video-game addicts and Internet pornography addicts. 'Cyberporn' accounts for about 25 per cent of Internet use and 25 million Americans (and probably 5 million Britons) visit porn sites on a weekly basis. An estimated 5 million people spend more than eleven hours a week viewing these sites, generating about $3 billion for the industry and ruining many people's lives. A combination of an addictive personality and genetic male programming can be a very dangerous one. Addicts can now join Sex Addicts Anonymous (SAA).

Living on the edge

F got his first motorbike when he was eighteen and crashed it within two months, fracturing his leg and causing his parents great anxiety. He had two more motorbike accidents before he stopped riding. He thrived on all kinds of dangerous sports such as caving, hang-gliding and bungee jumping and was banned for a year for driving too fast, even at the age of 35. He had also recently started to gamble, first on the lottery and then on sports results, and finally in casinos. He started to lose heavily and was in debt. In contrast, his sister had never had a motor accident and, although she liked sports, lacked his extreme risk-taking behaviour. She did enjoy gambling, however, but after a few costly trips to the racecourse knew that she, too, had a problem.

Injuries and fractures in young people are four times commoner in men than women. The number of car accidents involving men and women is similar in most countries but men are experts at the more dangerous and fatal high-speed crashes. Risk- or novelty-seeking behaviour is a genetic trait that is equally heritable in men and women, as shown by twin studies. A study for a recent BBC documentary asked identical twins to choose, independently of the other twin, a ride of varying thrills and danger in an amusement park. All the twins chose the same ride as their other half, including the roller coaster, which accelerated to 80 m.p.h. and did full circles. Despite their free choice, some twins found the experience so unpleasant that they were physically sick.

We've all inherited risk-taking genes from our ancestors, who had to take chances to survive – such as leaving a nice warm cave to find food. Why some people find risk taking particularly pleasurable is harder to fathom. Part of this may be the adrenaline and dopamine rush that is associated with it. Some of us need only a small excitement to satisfy our brains – such as buying a single lottery ticket once a month. Others, like F, have an insatiable appetite that requires them to feel sick with anxiety over the huge stakes involved in order to get a reward in the brain. We all get a greater feeling of excitement when standing outside in a lightning storm or playing roulette or the lottery than crossing the road. Yet the risks of crossing the road are far greater than being hit by lightning or winning the lottery (around 13–16 million to one in most systems). Modern humans are particularly poor about estimating our odds of success or failure.

The concept of estimating future odds arose in the seventeenth century from the work of Pascal and Fermat, which led to the launch of the insurance industry by Lloyd's in London. Previously, risk had been purely put down to fate and in the hand of the gods. The concept of odds of millions to one is hard for us to grasp, despite the age of super-computers. This may relate to our Pleistocene ancestors' universe and the experience of being confined to, at most, a hundred people in a group. Even when we are faced with simple odds like a 50–50 chance of winning by turning over cards – the highest card winning – our instincts betray us. If our opponent looks weak, insecure, unconfident and poorly dressed we bet much more than if the opposition is slick, confident and immaculate. Our brain reacts to this irrelevant information as if we were preparing to fight them to the death. Most of us vastly overestimate our risks of success: in surveys, 94 per cent of randomly selected men say they are in the top half of athletic ability. So it's no surprise that people expect to win the lottery or beat the casinos, despite the fact that Americans lose more than $60 billion annually on these miscalculations, and figures in Europe are similar. World-wide, gambling is still one of the biggest growth industries, with around 140 million regular Internet gamblers forecast by 2006. Although governments benefit from tax revenues from the winners, they pay out more on social welfare programmes for the losers.

Despite similar genetic influences, the amount and degrees of risk taking is generally much greater in men. The biological difference in the brains between the sexes appears to be not in the thrill or anticipation of the risk, but in the counter-balance of fear and consequences. Women have more brain activity and higher levels of the hormone serotonin in the fear centre (amygdala) than men – which in most situations restrains them from high-risk activities. This difference probably originates from the fact that men had to take calculated risks during hunting and, more importantly, women preferred the 'male' trait. Men also had to take major risks to compete with other males to show off their 'skills' in order to be more successful in attracting fertile females. The consequences of not competing were likely to be remaining a lonely bachelor and a genetic loser – so the stakes were high.

There are many examples of male risk taking from the animal world, such as guppy fish, which take it in turns to go in pairs and leave the school to scout for potential predators, so they can raise the alarm. Another is the Arabian babbler, a bird that deliberately tells a predator where he is hiding. These male birds behave like this only when females are around, and can show their skill in avoiding the danger they have placed themselves in. So why do females prize this rather senseless risky behaviour in males? Some of the attraction, it seems, is due to the signals it gives of their hunting skills and therefore ability to put meat on the table and feed their children – but this is not the whole story. It seems that the act itself is enough to signal their genetic potential for energy, strength and enthusiasm – all good signs for reproductive success, and good traits for any prospective children to carry on the line.

Keen observers have noticed that men are more likely to drive fast or to cross the road recklessly when women are watching. Females didn't need to take such risks to show off, as there were probably enough interested eligible men around as potential partners. In our ancestors it is unlikely that any females lacked mates. They also had greater responsibilities in having to collect the food to feed their children on a daily basis. The lesson is: be aware of your own primitive tendencies and try to substitute safer but equally exciting pastimes. It's worth realising that nowadays you don't have to take as many risks to survive or impress the opposite sex, and those who overdo it don't live long.

Greed and happiness

C always had everything he'd wanted. As a kid he was spoiled by his parents, who gave him the latest toys and games. He went to the best private schools and achieved good results. Rather than go to university, he went straight to the City and was a trader on the stock markets. He performed well in well-paid jobs, eventually leading after five years to six-figure bonuses. He bought Porsches and boats, had many girlfriends and drank champagne like orange juice. It was hard work and long hours. Some of his friends 'retired' but C was hungry for more. In 2001 the markets crashed, he lost most of his savings but kept working and spending as much as before. The slump continued and, like many, he lost his job. He turned to drink, then drugs and eventually committed suicide aged 29.

In the Western world we are acquiring more and more material wealth every year, with clothes, cars, jewellery and gadgets. In real terms income has increased nearly 50 per cent in the US and around 30 per cent in the UK and has doubled in Japan since 1970. Despite this, regular surveys over the last fifty years show that no country reports being happier or more satisfied with life. Greed for more seems to be endemic. We want more food, possessions, holidays, leisure, success, fame and wealth, believing that getting them will bring us happiness. Undoubtedly linked to this, depression and suicide rates are highest in the richest countries such as Scandinavia and Japan.

Why are we such a greedy race, never content with what we have? The answer lies as always with our ancestors and in the hormones and chemicals in our brains. We are genetically programmed to strive for more than we have. In our ancestors, who were constantly moving and couldn't accumulate possessions, this meant more food, spouses, status and children. All of these were important for the survival of our genes. Those ancestors who said, 'Don't worry about me – I've got quite enough' (food, status, wives, etc.) just didn't pass on many of their genes to us.

To give ourselves a short-term bribe for this greedy but genetically useful behaviour, we developed a reward system that gives us a sensation of pleasure when we accomplish something. The chemical involved is the ever-present dopamine, which gets released in a part of the central brain called the reward site (nucleus accumbens). Anticipation of a reward switches on dopamine receptors that trigger endorphins (our bodies' own opium-like chemicals) in another part of the brain, the hypothalamus. This produces feelings of wellbeing and stress reduction. Our bodies gradually become accustomed to this and so the system needs to be further gratified.

The system seems to work only when we continue to make progress. In other words, it is relative. We all know from personal experience that once you have achieved a goal you are no happier than when you started, but the *process* of achieving it does bring pleasure. This makes sense, and as a species keeps us all driving forward, striving for impossible success, rather than standing still like our early competitors the Neandathals, who died out because of a failure to move with the times. Unfortunately, being greedy in the modern world doesn't always equate with genetic success – Imelda Marcos's collection of three thousand pairs of shoes didn't help her offspring or genes – nor does eating vast amounts of food make you more likely to attract a mate or live longer.

Although we are driven by our genes to be greedy, like the greyhound that doesn't ever catch the mechanical rabbit it doesn't necessarily make us content or happy. Studies have shown that well-educated people are no happier than uneducated, rich no happier than poor, men no happier than women and religious people no happier than atheists. In other words, changes in our environment don't change our happiness other than in short term-gratification. A US twin study of several thousand middle-aged men and women found that self-reported wellbeing or 'happiness' was 80 per cent heritable. Socioeconomic status, educational attainment, family income, marital status or religious commitment could not account for more than a miserly 3 per cent of happiness.

This suggests that striving for an ultimate state of greater happiness by accumulating more is unlikely to be successful. The good news is that humans are optimistic and most of us rate ourselves as happy or very happy – in fact, most of us think we are happier than the average person. The happiest people have been found to be those who set themselves small, realistic, short-term goals, have close personal relation-ships, are committed to their work and who have low expectations. Buddhist meditators are one of the most relaxed and content of all humans. Scanning the brains of these monks has now shown marked increases in activity in the left prefrontal area of the brain – an area controlling positive emotions and good mood. There is also some evidence that they have less activity in the fear and stress centre (the amygdala). What is clear is that, whatever the mechanism, the more you actively strive for happiness, the less likely you are to reach it. It is, however, possible, with time and dedication, to reset our brains to be content with less and therefore have greater reward.

A matter of taste

M and his wife R were financially well off and suited in most things in life – except their taste buds. M liked salty spicy foods and curries and was always adding extra salt to his meals, despite warnings that this would increase his blood pressure. He liked sugar in his coffee and tried to substitute saccharin, which his wife couldn't stand. R was less keen on spicy and salty food, but liked her fruit and vegetables, being particularly fond of carrots. She was especially good at tasting and discriminating wines, and could tell straightaway if a bottle was off. They tried to bring up their children to eat a wide variety of foods and for the first four years were successful. After a while the kids became more fussy and difficult, with one refusing to eat fish and the other anything green. M and R often wondered whether their different backgrounds had shaped their own taste differences – but could nature also play a role in taste?

One of the early pointers that taste may be genetic came from reports in the 1930s that two-thirds of people detected a chemical called PTC – used in minute quantities in ice production – as tasting very bitter, and thought it inedible. One in three couldn't detect it at all. Individuals vary widely in their perception of bitterness: 60 per cent of Asian Indians, 66 per cent of Europeans, 90 per cent of Southeast Asians and 97 per cent of West Africans perceive the chemical as bitter, whereas a minority of individuals find it tasteless. Taster status is higher in women than in men and a group with even more enhanced tasting abilities have been identified called 'super-tasters'. Because they are so sensitive, these super-tasters often have an aversion for bitter-tasting foodstuffs: beer, caffeine, artificial sweeteners such as saccharin, and thiourea-containing vegetables, such as broccoli, Brussels sprouts, spinach, cabbage and turnips. Super-tasters also appear to use less sugar and fat in their daily diets and on average are thinner than nontasters, who, in contrast, may need more fat and sugar to get the same 'buzz'.

Simple tests on our twins have shown that even a crude preference for salty or sweet foods is variable and at least 50 per cent genetic, with the remainder being due to upbringing and experience. A test of chocolate preferences found, however, that preferring milk or plain chocolate was entirely due to upbringing and presumably exposure to either sort at an early age. However, your taste and texture genes mainly determine whether you prefer hard or soft centres in your chocolates.

Overall, men generally prefer saltier, more bitter tastes and women sweeter ones. This may be due to the different roles of our ancestors in food catching, with females tasting and gathering a wide range of berries and fruit, some of which may not have been ripe or eatable, and men killing a smaller range of edible animals, but perhaps needing more salt during long periods of hunting without water. Females have ended up on average with a much more refined palate than men, as well as being able to discern subtle differences in sweet and sugary tastes.

Why we have such a wide variety of food preferences and taste receptors is unclear. Perhaps there was an advantage to individuals in a small tribe if not everyone ate the same foods. This may have avoided food fights, helped tribal members to adapt to new foods as the tribe moved and possibly prevented the whole tribe being wiped out by food poisoning. Certain foods can be poisonous to some people and harmless to others. A good example is the eating of undercooked broad beans, which are toxic only to people with a gene that protects them from malaria.

Scientists have discovered around ten thousand taste receptors on our tongues to detect sweet, salt, bitter and sour and possibly fat. Each taste bud contains around 100 receptor cells carrying G-protein-coupled receptors (which are related to receptors for pheromones). When these receptors encounter a piece of food, they bind to the food, recognise the taste, and switch the cells 'on' by prompting them into an active state. The cells then transmit information to nerve cells that relay the data to the taste centres of the brain cortex. In humans, the gene controlling bitter taste has just been found by a study of large Mormon families in Utah. Researchers found the gene to be on chromosome 7 and part of the TAS2R bitter-taste receptor family. They called it the PTC gene. It has five main variants and these are found in all human populations. All human groups tested include some individuals with the super-taster gene, and most populations have individuals with nontaster genes. The exception is American Indians living in the southwest part of the US, suggesting that their increased ability to taste was crucial for survival. Perhaps bitter foods and drinks will in the future have two different strengths for people with and without the PTC gene.

Junk-food junkies

V was a cab driver who enjoyed his job and often worked long unsocial hours. Lunch and sometimes dinner was always on the move – usually in a fast-food restaurant. Recently, so he could stay in the car, he'd taken to using the drive-through establishments that were appearing everywhere. Five years after starting the job, he realised his waistline was definitely increasing faster than his age. A short while afterwards he attended an insurance medical and was told he had major problems – high blood pressure, obesity and early diabetes. His doctor told him he needed to cut out all the fats in his diet. He managed to switch to diet Cokes and reduce his beer intake, and for a week or so resisted the burgers and pizzas in

favour of salads – but found it hard going. After two weeks he
found himself fantasising about fatty meals. Despite the guilt,
the burger and fries the next day tasted really good.

Many common diseases are precipitated by our over-
indulgence in a few specific foods that we naturally crave.
They were always in short supply for our ancestors, who were
selected to crave them for their survival, and they passed this
trait on to us. These rare items included salt, which helped
prevent dehydration but now leads to high blood pressure;
fats which had the highest energy value for storage; and
sugars for short-term energy. But fat and sugars lead to heart
disease and obesity. These natural cravings are further
encouraged by the food industry, which in the US alone
spends $32 billion a year telling us how much we need them.
These willing consumers spend $160 billion a year on fast
foods, accounting for half the average family's food budget.
The industry also tricks our taste buds. It routinely adds beef
flavouring to 'vegetarian' fries, and barbecue flavouring to
blander foods to excite our palates and genes.

Our bodies don't stand much of a chance. In the ten
minutes it takes to eat a typical 'fast meal' of burger, fries and
milkshake , we obtain all the calories we need for the day and
125 per cent of our fat requirements. This sudden hit of fat and
calories causes a surge in our blood-sugar and levels of leptin,
a circulating hormone produced from fat cells. This triggers
the release of endophins (morphine-like substances) in the
brain, which in turn causes dopamine to be released in an
area called the nucleus accumbens – resulting in a pleasing
reward sensation. The similarities to drug and alcohol
addiction are obvious. In experimental animals, overfed rats
developed progressively higher leptin levels and needed even
greater fat intakes to get the fat cells to produce them. As with
a drug, the more fat you eat the more you need it to sustain
your pleasure levels. Overfed animals get classic withdrawal
symptoms when the high-calorie food supply is shut off
rapidly – so-called 'cold turkey' (without the mayo and fries).

Like all addictions only a proportion of us have major problems, but unfortunately we can't identify these individuals yet.

Studies of our ancestors in the days before they became farmers reveal that they had amazingly varied diets, consuming more than three hundred types of edible plant and animal life, mainly fruits and berries, enriched with the occasional protein meal in the form of fish or meat around once or twice a week. Compare this with the paltry number of basic items eaten regularly in most modern Western diets. Ten core foods now account for 80 per cent of the world's current eating habits: wheat, corn, rice, barley, sorghum, sugar, potato, sweet potato, soybean and banana. Our palates and genes evolved to cope with the rich variety of tastes and foods available and we are one of the most flexible eaters on the planet. Our bodies would be much healthier if we returned to a diet closer to those of our ancestors with more variety and less of the previously rare foods we now crave in abundance.

Most people don't realise that all our modern foods are the product of a form of natural genetic modification (breeding) by our ancestors four to thirteen thousand years ago. In a typical fast-food meal, V will ingest the descendents of animals and plants that our clever ancestors discovered had potential as foodstuffs. In animals, this potential came from rare characteristics such as the ability to be domesticated, displayed by wild aurochs (beef) from the Middle East or jungle fowl (chicken) from China. In plants, mutant varieties that didn't reproduce normally helped farming. Examples include mutant corn from Mexico, the seeds of which wouldn't germinate normally; mutant peas from Jordan that stayed in the pod rather than spontaneously bursting; and mutated, seedless, sterile bananas. Many of these foods are easy to cultivate and eat, but like the banana, are at current risk of extinction becuase they cannot defend themselves against disease. Until recently, genetic manipulation of our food has been by very slow gradual refinement. This has been by artificial crossing of strains of hardy varieties with nutritious ones to increase harvests and yields.

GM may change this. Proponents promise us fruit that doesn't go mouldy, crops that don't need insecticides, rice with essential vitamins, beans that are nonallergenic, vegetarian fish and eye-friendly onions. They tell us that in twenty years most of the world's oceans will be overfished and that GM fish farms will be the only way to supply the world. No one knows whether the potential environmental risks outweigh the benefits and, for some reason, people don't completely trust the food industry.

Revenge against fast food may be coming. In July 2002 a 56-year-old called Mr Barber was the first to sue McDonald's, Burger King, KFC and Wendy in the US for their share of the blame for his heart condition and diabetes and a failure to issue warning notices. Perhaps all fast-food joints will be forced in the future to issue health warnings, and it may become illegal to sell a burger to a minor or someone with the wrong genes – burger prohibition.

Occupational hazards

When F was ten he wanted to be like his dad and grandad – a medical doctor. However, over the years he started to resent the parental pressure expecting him to follow suit. Like any normal teenager, he didn't want to do as his parents wished and opted to study computer science, at which he was skilled. At the age of 21, although he'd graduated well, he now felt he should have studied medicine, which would have suited his social skills more. It was too late to do this in his own country. He successfully studied abroad, eventually qualifying at the age of thirty – happy but with considerable debts. Looking back, he often wondered whether his home environment or his genes had pushed him into medicine.

As personality (introversion, extroversion, openness, neuroticism etc.) has a considerable genetic component, it's not surprising that job selection is also partly genetic. Twin studies of rare sets of identical pairs separated at birth and raised apart have shown that they can often end up in similar jobs. One example is a male pair who, having lived with foster parents, eventually met up in their forties for the first time and found they were both working as firemen. A set of triplets was brought up believing they were fraternal twins (sharing only half their genes). They all obtained masters degrees in the related areas of nursing, social work and special education and found out later that they were, in fact, genetically identical. Not all identical twins pick the same types of job – one male twin became a doctor and his brother started as a medical scientist, but changed to become a graphic designer. His brother later admitted he would rather have been an artist, too. Overall, two-thirds of identical male twins said they liked their co-twins' choice of job and only one in ten expressed a dislike.

The example used of medicine is a good one. Studies of candidates for places at medical school are eighty times more likely to have a medical relative than students who don't apply. Obviously, some of this selection is due to the strong family pressures as well as the genes. Some people choose jobs on the basis of their hobbies and pastimes. Usually, people have more choice over their leisure activities than their actual occupations and these may better reflect their real preferences. Twin studies examining normal and reared-apart twins for their similarities in interests and aptitudes have shown that these, too, are genetically influenced, with heritabilities ranging from 40–70 per cent for areas such as practical skills, arts and crafts, travel, buying and selling, and socialising. Some identical twins raised apart found, when they met for the first time in middle age, that they shared special hobbies such as woodworking – a hobby group that had one of the highest heritabilities at 76 per cent.

An added complication to job choice is that of gender. Although subjects such as medicine are nowadays equally represented by males and females, most other occupations have strong sex biases. These partly relate to the various innate individual skills of males and females. Although there are always exceptions, males predominate in jobs that involve spatiotemporal, map-reading and higher-maths skills, which were presumably originally useful for hunting. These include architecture, engineering and flying aircraft. Women are most successful in jobs that require networking, communication and fine hand coordination skills, which were useful for fruit gathering, child rearing and socialising. These jobs are to be found in the fields of literature, linguistics, administrative, and organisational work and teaching. Women probably have better memories for fixed spatial objects, which would have allowed them to remember the site of special fruits or berries from previous years.

Evolutionary psychologists have long argued whether a woman's place is historically in the home or in the workplace. Evidence from different cultures and primitive tribes is that, to some degree, all women continue working and maintain their own independence. The recent phenomenon (in the last fifty years) of permanent dependent housewives in the affluent West doesn't appear to be part of the evolutionary plan. When choosing a career path, whether you're more likely to be successful following your genetic instincts is an unanswered question. This strategy relies on whether your parents made the choices that suited them. Many great men and women decided to ignore their families and strike out a new path and destiny for themselves.

Fears and phobias

M was a confident forty-year-old professional in charge of twenty people. He was giving a departmental seminar in his normally calm, authoritative and confident way. Halfway through his talk, he glanced to his left and less than three feet away from him on the wall saw the creature he feared most – a spider. He stopped dead in his tracks, and was literally paralysed with fear. His mouth was dry and he couldn't speak. He turned white and sweated profusely. His pulse was racing at 150 beats per minute, and his stomach tightened as blood drained from it. After twenty seconds, which seemed like an eternity, he was able to move and he made for the door. His audience was initially stunned and then started to laugh. How could a six-foot man be so scared of a 1-cm spider?

Our totally fearless and therefore reckless ancestors probably didn't last many generations. They were quickly squashed by a mammoth or eaten by a sabre-toothed tiger. Those who were overfriendly with spiders or snakes also probably didn't get to live to pass on their genes. Our fear response is essential to our human survival. The key changes in cortisol, adrenaline, blood flow and heart rate all happen within seconds, priming us for a fight or to run away (a flight). We all have imprinted in us a fear of crawling or slithering, potentially poisonous animals. The strength of this varies between individuals. Small babies will flinch when a plastic snake is placed in their cot, but will happily play with a loaded gun or hand grenade. These inbuilt fears are irrespective of where you live: fear of snakes also occurs in countries such as Ireland, where there are no natural snakes. These fears can be modified to some extent by early learning or conditioning, but it is difficult to train animals or people that snakes are harmless or that they should be frightened of flowers.

This innate genetic fear is probably further influenced by early learning (conditioning), our surroundings and parental warnings and reactions of others. Some individuals have inherited grossly exaggerated fears, which, when out of proportion, are called phobias. The fear response is a very basic one and occurs in a primitive part of the brain the size of an almond, called the amygdala. We are conditioned to worry more about 'natural' risks than modern ones. We are all naturally fearful of storms, lightning, earthquakes, deep water, predators and childbirth. This is despite the statistically minuscule chances that these events will kill or injure us in the modern age, compared with more common and certain events, such as disease, firearms and road-traffic accidents.

Fear conditioning has been tested in twins in some stressful situations involving spiders and electric shocks. There was considerable variation in the speed of learning, the response (fear and sweating) and how quickly fear subsided. All these variables had a heritability of 35 to 45 per cent, showing that even 'learned' behaviour was a genetic component.

One of the key genes for this fear reaction has now been identified: the serotonin transporter gene (5–HTT). In an experiment using special (functional MRI) brain scans to assess the responses of individuals when shown pictures of people experiencing fear and distress, responses differed for different genes. The right side of the amygdala in people with the 's' form of the gene (the fearful form) lit up like a beacon when stimulated this way, compared with those with the common forms of the gene. Their brain function was otherwise identical and normal. The 'fear' genes probably work (when faced with triggering stimuli) by allowing warning chemicals to flood the amygdala at a lower threshold than normal. This produces the primitive protective response with all the well-known side effects.

These neural amygdala connections are particularly extensive and rapid to the eyes and ears, so that we are built to feel danger before we get to think or reason about it. All of us have this rapid fear response built into our systems; it is usually protective, making sure our bodies are ready to fight for our survival within a few seconds of a threat. It is quite easy to see how this finely tuned but basic survival mechanism, designed for us perhaps thousands or millions of years ago, may be an oversensitive burden in a few individuals. The genes of anxious or phobic people may have been ideal for survival in protecting them against deadly 10-metre snakes and dawn raids by predators or other tribes, but are a hindrance in the suburban idyll of the modern world, where predators and reptiles are of a different and subtler nature.

The sucker gene

V was often disappointed in people – they tended to let her down. Ever since she could remember she had put trust and faith in those around her, and they had usually reciprocated. V loved company and having people around her. She was always sure to send all her friends thank-you notes and birthday cards. Sometimes, however, things went wrong – like the time she lent the au pair girl a large sum of money to fly back to Australia and never saw her or the money again. Her husband, who was much more hard-nosed, said she was too kind for her own good and sometimes called her a sucker. But, however hard she tried, she always saw the good and not the bad side of people. She was like her mother in this respect and wondered whether it was her upbringing.

Many of us are overgenerous and gullible in some situations, and others are the opposite: crafty and streetwise. This part of our personality is partly genetic and partly environmental. This behaviour is all relative – most of us are very wary of giving credit cards or car keys to strangers. But why should we be so different in our ability to trust each other? Humans are the most social of animals and depend on a high level of cooperation and collaboration to survive effectively. Our ancestors' instincts on trust were crucial in areas such as lending food in times of famine – for example, to a starving companion – in the hope that he or she would repay the debt when they themselves fell on hard times. If they got it right and the person was trustworthy, the gift was repaid. If they got conned the price might have been death. Studies of the human brain show us that some of its largest areas are involved in recognising subtle facial patterns of fellow humans – clues to their trustworthiness – a skill regulated by genes.

A good example of trust genes in action is found in vampire bats, who live in caves in their thousands. Each night these bats go out and get a takeaway meal of high-protein blood from other animals – but some return empty-handed and go hungry. Although they will look after their kin first, some will regurgitate blood to feed nonrelatives who are hungry. Over time, researchers found that bats gave only to those that had themselves given in the past – and refused the double dealers. To do this they needed a very good memory to distinguish the good and bad guys from the thousands of companions – and indeed they have one.

So if trust is so crucial, how have patsies like V survived all this time? The answer is probably that suckers will on the whole have certain advantages, such as having a wider social circle – useful for support in times of trouble. Being generous pays off when there is a large social network where you are likely to re-encounter people and when the benefits of occasionally receiving rewards back outweigh the costs of giving. For example, this might work when there is an occasional threat of starvation. The stereotypically shrewd individual will trust fewer people and have a much smaller support network and will benefit most when he/she doesn't have to see people again and reciprocate. The other good reason to be a bit of a sucker is that kindness, understanding and generosity are top of our instinctive shopping list when we pick a long-term mate. Sly careful humans may not give much away – but they're not regarded as attractive characters.

In computer-game theory, the game of Prisoners' Dilemma is often used as a model of how human evolution has moulded our behaviours and social strategies. In this game noncooperation (defection) by one player provides a short-term gain to the noncooperator. If both cooperate, both win a small amount. But, if both defect, both lose out. Many simulations and games have shown that the best strategies involve being initially generous (sharing), but reacting badly if double-crossed – the so-called tit-for-tat strategy. Studies have shown that humans are very good at detecting cheats from their facial clues. A group of students correctly identified defectors in the game of Prisoners' Dilemma on the basis of their photographs – and thought they looked familiar, despite never having met them.

Finding the right balance is the key. We know that a few key facial patterns can make a difference to our ability to trust someone – smiling is the most important, followed by long eye contact and dilated pupils. Our instincts towards trust are constantly being challenged by the tricks employed by smiling bright-eyed actors in commercials, but, once we've been ripped off, we would never repeat the episode with the same person or company – our long memories make sure of that.

Heroes and cowards

Two teachers, C and F, were taking a group of thirteen-year-old boys, including C's son, on a three-day river-canoeing trip. There were two boys in each Canadian canoe. The first two days had been in glorious conditions; however, overnight, while they were camping, it rained heavily. In the morning of the final day, the water level had risen nearly 2 feet (60 cm) and the river was moving faster. The morning's paddling was uneventful. As they approached the most risky part of the river, a Grade 4 rapid, they discussed whether to continue on foot or by river. They decided to proceed cautiously, one boat at a time, with the teachers going first. The teachers navigated the first stage and waited in an eddy. The first boat came down containing C's son and another boy. They unexpectedly hit a rock and flipped over and continued down the rapids below the teachers. Without a moment's thought, C dived in after them. He got to his son, who was stunned, and managed to get him to hold on to a branch. The other boy was already drifting further down, and he went after him through the raging rapids. Neither was ever found again. C was called a hero and F a coward, who was ostracised and lost his job and status.

We all react differently in certain life-or-death situations. There are many instances of individual heroism in humans where one individual risks death to save another unrelated human – and we are the only species to do this. If we are but slaves to our selfish genes, why should we sacrifice our lives? The explanation of why a father would instinctively risk his life to save his son is easier. His son carries 50 per cent of his genes and is his lifeline to immortality. Not all fathers would necessarily sacrifice themselves for one child, because they could still in theory have many more sons and daughters, which would not be the case if they died.

Why humans occasionally risk death to save a stranger is more difficult to explain. Is this real altruistic behaviour with no benefit for the individual? Although the reactions may be instinctive – and they may not think about the consequences at the time – the rewards if they survive are great: adulation, medals and markedly increased status (good for survival and reproduction) can often follow in societies that encourage such actions.

Heroic deeds by men, although usually instinctive, more often involve saving women than men, and are also more likely when other females are watching. These deeds are likely to increase future mating potential, as the traits are attractive to women. The other benefit is that do-gooders naturally feel good about themselves, receiving a natural high induced by chemicals (dopamine) in the brain. Twin studies in the UK demonstrated that self-reported altruistic behaviour or intentions was a strongly heritable trait – with a heritability of around 60 per cent. Levels of altruism increased with age, and were on average higher in women than men.

Instinctively altruistic (or, as we discussed previously, trusting) behaviour may have evolved from a tribal need in humans to cooperate for mutual advantage, expecting the act to be reciprocated later. The other possibility is that it evolved because we lived in small groups whereby most people were related and were therefore likely to have carried shared genes (kinship theory). Other people are naturally less altruistic (natural cowards): they prefer self-preservation and the possible negative social consequences to the potential downsides of death or injury and losing their genes. Human society is made up of a full range of instinctive heroes and cowards, and many shades in between. Each of us has inherited his or her own personal strategy to survive, and in the future we may discover related genes for heroism and cowardice which may confer hidden advantages or disadvantages.

7

Minor Body Irritations

As we get older we start to see our bodies' imperfections more clearly. Many natural differences between us tend to show themselves – some of these are good, others bad. Many of these variations are genetic effects that are triggered at certain ages or by certain environments. Some traits may now be disadvantageous in the current environment that were previously advantageous – either in our early life or in those of our ancestors.

A pound of flesh

B was always on the plump-comfortable side of the scales but, after she had had two children and reached forty, there was no doubt her weight started, slowly but steadily, to increase. She went on various special diets that helped control things for a short time, but she usually relapsed. She was careful in what she ate, generally avoided sugars and snacks and, when she calculated it, found that her calorie intake was actually below average. Her weight continued to increase slowly. Although she said she was following a diet plan from a recommended book, her family and friends didn't seem to believe her. She felt bad that no one trusted her and soon gave up the diets, became depressed and became even fatter.

Obesity and the ingredients that cause it are all strongly genetic. Many studies of twins and families show that 60–70 per cent of the differences between people are determined by genes. Obesity is the major modern epidemic affecting all developed countries, with rates of associated serious disease rising everywhere. The US is leading the way: more than one in four children and one in three adults are clinically obese, leading to massive risks of heart disease, diabetes and arthritis and many other health problems. Just in the last decade, rates of obesity in the US have increased by one-third. During this same interval, the weight of the average American increased by nearly 9 pounds (approximately 4 kg). The UK and other Western countries are not far behind and are catching up quickly.

Doctors define obesity in a complicated metric way called body-mass index (BMI). A BMI of 20–25 is desirable, 25–30 is overweight, over 30 is obese and over 35 severely obese. You calculate it by taking your weight in kilograms and dividing it by your height in metres squared. Someone 5 feet 4 inches (1.6 m) tall would be obese if over 79 kilos (174 lb or 12 stone 6 lb). If taller (say 5 feet 10 inches, or about 1.8 m), the cut-off for obesity is higher, at 95 kilos (209 lb, or 14 stone 14 lb). The number of severely obese people is also increasing dramatically in most countries. As B found out in the above scenario, women put on more weight between the ages of forty and fifty than at any other time.

The causes of obesity are simple. Energy intake (food) exceeds energy expenditure. We are today surrounded by an abundance of available food and calories. The US food industry now produces enough food to supply twice the US population's daily requirements and succeeds in selling most of it. The mechanisms underlying obesity are much more complex, and a number of organs (such as the brain, intestines and liver) and many different mechanisms are involved. There are short-term and long-term mechanisms that regulate the fat on our bodies. Messages to eat or to stop eating are received in the specialised feeding centre of the brain (the arcuate nucleus), which in turn alters behaviour and energy expenditure. The long-term eating-regulation messages come mainly from a hormone called leptin, which is released in increased amounts when fat stores rise, and increases energy expenditure. Leptin levels and energy expenditure are decreased when fat levels drop. This controls the amount of fat that is stored long-term, working like a thermostat. This sounds like a perfect system. The problem is that the leptin-regulating system always selects the historically highest fat levels as the baseline, which it regards as normal. In this way, if one winter you put on weight and reached 80 kilos, or 176 lb, any drop below that, even a year later, will be seen as evidence of starvation. The system is considerably more sensitive and effective in preventing weight *loss* than weight *gain*, on which the hormone has only a modest effect.

Short-term regulation of eating habits and hunger come from a variety of other hormones (such as ghrelin and PPY) that affect the same part of the brain (the arcuate nucleus) and are released by the intestines and liver after a meal. The more people diet, the more ghrelin is produced, causing more hunger. Very obese individuals have abnormally high levels of this hormone, as they become immune to its effects. PPY acts in the opposite way and stops hunger. All these different hormones and delicate control mechanisms vary between different people and are all genetically influenced.

A recent twin study showed identical twins to have very similar eating habits, such as preferences for the number of times they ate in a day, as well as the particular times they felt most hungry and even the number of eating companions they preferred. By contrast nonidentical twins showed no clear similarities, suggesting that upbringing has little effect on eating habits. Obese mothers are often accused of over-feeding their kids, and that is one reason why the disorder is familial. However, in a twin study, overweight mothers were no more likely than normal-weight mothers to offer food to deal with emotional distress, as a form of reward, or to encourage overeating.

The genes for each of these regulating processes all have a role in adult obesity. This means that everyone has genetically programmed levels of appetite, will respond to different foods at different times of day that make them full at different thresholds, and will process the food differently, producing energy or fat in different proportions. Dietary guidelines say that average calorie levels needed are around fourteen to twenty calories per pound of body weight, depending on exercise levels. For women of average size and activity, this is around 2,000 calories – for men, around 2,500 calories. However, these 'average' levels are fairly meaningless, because the effects of everyone's genes on the regulation mechanisms at different ages are very different. This explains why B in the scenario, who has a low metabolic rate, puts on weight at calorie intakes at which others would easily *lose* weight.

A number of specific genes influencing obesity have now been identified, including the genes influencing the following hormones and enzymes: leptin, PPAR ADRB3, CCK, MC3R and UCP. These are useful to explain problems in rare families with extreme obesity problems, although are not yet diagnostically good enough for the commonest obesity problems. Until then, many of these genes are proving useful in designing promising future therapies.

Obesity is a good example of how previously useful genes respond to an altered environment. This is clearly seen in recent history, as when Pacific Islanders or south Asians move to modern affluent Western diets and begin to suffer from obesity, heart disease and diabetes within a few generations, at similar rates to the original population. The same genes that now predispose us to obesity and its health problems undoubtedly saved our ancestors from extinction, because having extra stores of fat and lower metabolic rates allowed them to survive in times of semistarvation longer than others. A good example of this is modern-day Samoans, who with other Polynesians have the greatest propensity to obesity. These people survived epic sea voyages in primitive crafts thousands of years ago. Only those with the best metabolism can have survived to eventually populate the islands. Modern man, if healthy, can survive on water alone for about two months, as evidenced by IRA hunger strikers. Those with slower metabolism and more fat deposits would lose muscle less quickly and remain stronger. Well-insulated humans could also survive longer in cold water or air than their skinnier and faster-metabolising compatriots, as demonstrated by the survivors of naval disasters. These survivors are our ancestors.

Nowadays, when people predisposed to obesity go on diets and start to eat less, after a loss of only a pound or a few ounces – their bodies respond as if in a crisis of famine. Their metabolic rate, which is already a bit slow, reduces even further to prevent calorie and fat loss and hormones are released to stimulate hunger. Thus, paradoxically, dieting can often have the opposite effect. When diets do work it is usually because the regime is so complicated that dieters spend hours working out what and exactly when to do their eating. Low fat or low carb diets (such as the Atkins diet) do produce weight reduction in the first few months, but most are unsustainable and have side effects such as constipation and kidney stones. In most studies of dieting drugs, the dummy or placebo tablets have nearly as much success as the real ones. Attempts to fool the body with items like artificial sweeteners don't always work. In trials where saccharine has been secretly substituted for sugar, subjects ended up eating more calories to make up for it.

In contrast, increasing levels of exercise will increase the metabolic rate and cause loss of fat. This is why exercise is a more efficient long-term method of losing weight than dieting. As obesity still carries an enormous social stigma, with major discrimination in many walks of life, it's about time we started treating sufferers seriously and with as much sympathy as any other genetic disease – for without our cold- and famine-resistant ancestors we probably wouldn't have made it past the last ice age.

My bum's too big

D was pretty and energetic, with a good appetite. She was not overweight and yet she constantly worried about not being able to shift the fat from her bottom and thighs, despite sessions at the gym three times per week. She always felt she wasn't doing enough exercise compared with some of her svelte friends. She developed a bit of a complex about her body shape and tried all kinds of diets and short-term exercises unsuccessfully.

Although it is reasonably well known that obesity is 70 per cent genetic, many people are unaware that body shape and the distribution of body fat is equally determined by their genes. This has been shown by our twin studies, which demonstrated a genetic influence on shape that was independent of how fat they were. We found that the amount of fat stored on women's thighs, bottoms and bellies is all controlled by genes – with heritabilities of around 60 per cent. Through evolution, women collected more fat than males to help them through the extra demands of childbirth and breast-feeding. The extra fat deposited particularly on bottoms and thighs was useful as a survival mechanism in times of famine. This was because fat in these areas, well away from major blood supply and organs, was less likely to be used up for routine energy requirements and was stored until really needed. It was probably also useful when early humans migrated out of Africa into colder areas of the world. Modern research has shown that women with the classical pear-shaped figure (like Marilyn Monroe) have less heart disease and diabetes and live longer than men or women with more rounded, apple-shaped figures.

The other likely reason for female curves is that our male ancestors have always shown a preference for women with this shape. This is seen now in recent surveys of diverse cultures (both primitive and modern) throughout the world where men prefer a certain shapely figure (a waist-to-hip ratio of 70 per cent – e.g. 24-inch waist and 34-inch hips), regardless of the total weight. Dedicated researchers performed a time-consuming and arduous survey of the vital statistics of 577 *Playboy* magazine centrefolds over the last fifty years, which confirmed our overall preference for this magic ratio. However, it also showed that male preferences of the ideal shape for a sexy figure have changed slightly with time. Although the weight of models remained the same, their height increased slightly so that the index of fatness (Body-Mass Index) has dropped from an average of 20 to around 18. In the 1950s and 1960s busts of models were bigger (and natural), waists smaller and hips bigger with the resulting waist-to-hip ratio being more extreme at 65 per cent compared with 70 per cent today.

If the average *Playboy* reader is typical, modern men are now showing a slight preference for more androgynous figures with smaller hips and breasts. We've previously seen that men prefer curves that are indicators of health, fertility and a nonpregnant state. The male preference (and some would say obsession) for breasts is more of a mystery, as having large or fatty breasts doesn't seem to provide any greater efficiency in breast-feeding. Female apes don't have obvious bulging mammary glands until they start lactating. Men might have originally selected women with clearly defined breasts as a general sign of good health and body symmetry or just because one successful group of men liked them.

Long-term studies have shown that a BMI of 20 and a waist-to-hip ratio of 70 per cent give the optimal chances of longevity, health and fertility. So current trends that see men showing preferences for skinnier women may not be good for either sex. Fat on the female bottom is a much healthier sight than the male form of fat deposits (the beer belly or paunch). This may have evolved as the best place to store fat in men and still give them a chance of running after their perfect girl and their dinner, but nowadays it leads to an increased risk of heart disease and diabetes.

The message for women like V in our scenario is that, however much weight you lose, the proportions of fat distribution tend to stay the same – so, unless you started with a lot of fat on your belly, you'll tend to maintain the same overall shape. Mothers' and daughters' shapes are often very similar. The response of every individual body to work-outs in the gym will vary in the amount of fat or muscle that is altered. Every individual has different dietary, exercise and calorie needs. When you look at your figure on the beach, current cultural pressures may make you less than totally satisfied. However, you should be thankful for them: the fat deposits your genes have given you were essential during the last ice age in keeping you alive and attractive and even today keep you healthier than men.

Chilly fingers

H and her husband M were always fighting over the central-heating thermostat controls: she was always cold and he always felt hot. She would often complain of cold hands and feet and disliked cold weather. H also refused to go on skiing holidays, which annoyed him, as he felt the temperature was just an excuse. She preferred hot sunny holidays, but he felt too hot and sweaty and became irritable. Someone was always miserable on holiday. Why were H and M so different?

External body temperature varies between races by several degrees and is strongly genetically controlled. In experiments in which Africans and northern Europeans immerse their hands in a cold-water bath for one minute, the differences are more marked. The blood vessels of Africans react vigorously and shrink, whereas European blood vessels hardly change. As well as affecting the skin, these changes could explain why Africans in cold climes are more prone to high blood pressure. A number of chemicals and hormones (the adrenergic system, endothelin, angiotensin) control how blood vessels shrink and these are all under the control of genes.

Cold extremities are due to a shut-down of the blood vessels nearest the skin, which can go into spasm. It is related to the lay term 'chilblains' and, when they cause pain and change in colour of the hands or feet, it is called Raynaud's syndrome. Our twin studies have shown that the responses to a 'cold challenge', as the test is called, are quite variable within Europeans. Some subjects' hands go white and painful; others feel nothing. These changes were 60 per cent heritable. Studies of women with Raynaud's whose hands turn white or blue in the cold also showed a strong genetic influence.

Modern man came out of Africa around fifty thousand years ago, displacing the tough but non-evolving Neanderthals, who hadn't made it further north than Germany. As they spread outwards and northwards, and diversified, the genes of northern races slowly adapted to the harsher climates, allowing their blood vessels not to go into spasm, and continue to warm their bodies. They were also more innovative than Neanderthals, using sophisticated tools for fishing and hunting, building houses and sewing clothes – all useful in the cold. We now believe that clothes have been used by humans for around 60,000 years. This estimate comes from studying the genetic changes in body lice who migrated from our heads to our clothes at this time. Eskimos adapted to withstand for hours cold temperatures that would kill others within minutes. Regulating our internal temperature is very crucial for humans, as the internal system breaks down if the blood temperature changes by a few degrees – as seen when we get fevers – and changes of 5°C can be fatal. The regulation of our external body temperature occurs by changing the blood supply to the skin. Australian Aborigines, who had fifty thousand years to adapt to a desert environment, dramatically reduce their skin temperature and heat loss at night in the cool desert, conserving energy and reducing shivering.

Keeping cool or warm is a reason why body shape in humans has evolved differently around the world. Those in colder climes tend to have a greater body-mass-to-surface-area ratio, so they are stockier with shorter limbs, and those in tropical areas are thinner with longer limbs to keep cool. We all contain genes originating from our ancestors, who endured the tough African sun. Some of us will have more of the mutated genes that enabled us to travel to cold climates and survive. It's the mixture that's important. So, if you're living in Alaska or Scotland and have the misfortune to be tall and skinny with cold blue hands, blame it on your ancestors, who should have migrated to Florida instead.

Bugs and bites

K and her sister lived in a humid hot part of Florida. They looked pretty similar, but throughout their life they had both noticed that it was K who always got bitten by mosquitoes, which had always seemed to leave her sister alone. K remembers several times while camping waking up in the morning covered in bites around her head and feet, with large itchy weals, which made her face puff up. K refused to go camping and wanted to move home further north, where she could go outdoors. Neither she nor her sister could explain the differences between them.

Mosquitoes have been on the planet for much longer than we have: around 250 million years, the oldest known specimen being 79 million years old. They have evolved and adapted to be one of the most successful species on earth, with at least three thousand different varieties, surviving in the tropics, the Himalayas and the Arctic. They have adapted to breed in tiny drops of water in man-made environments, such as those found in old tyres. They cleverly pierce the skin of animals they feed on with two sharp tubes – one sucks the blood and the other injects a mixture of saliva, anaesthetic and an anticlotting factor. When the anaesthetic wears off, the saliva leaves a nasty allergic reaction, which swells and is itchy.

The varieties of mosquito that drink human blood, unfortunately, also pass on parasites that cause malaria, yellow fever, dengue fever, filariasis and other nasty infections that kill 3 million people a year. Attempts to eradicate malaria in Africa with pesticides and treatments have largely failed, as the mosquitoes have simply evolved new genes to produce enzymes to make them resistant. They breed fast and so can mutate quickly in human time frames. Recently, hope for freeing the world of these diseases has come from genetics. The mosquito genome (*A gambiae*) has been fully sequenced, leading to the possibility of producing genetically modified mosquitoes that don't like humans or can't carry the malaria parasite.

Finding out about the genes has also unlocked some of the secrets of their success. Mosquitoes have very keenly adapted smell receptors controlled by about three hundred genes. Females (the only ones that bite humans) mainly attack at night and are attracted by body heat and hot breath. They prefer areas of the body that have a good blood supply, such as hands and feet and joints. Strangely, they seem to home in on certain body odours: some prefer sweaty lactic acid smells, others ammonia smells. The common species of mosquito (*A gambiae*) favours attacking the foot because of the smelly, cheesy odour of the feet of some people. In fact they are also attracted to Limburger, a particularly smelly Dutch cheese that contains a similar bacterium (*Brevibacetium*) to the one that lives between your toes and gives the characteristic bouquet when feet get hot.

Other animals have developed genetic mechanisms to protect themselves against malaria. The giraffe is known to give off a particularly pungent odour that can be smelled 250 metres away and the males are known as 'stink bulls'. Their odour – a cocktail of around ten different chemicals – protects them against bacterial infections, fungi (such as athlete's foot) and mosquitoes and other blood suckers. Nobody likes the smell, except for female giraffes, who find the pungent aroma sexually stimulating (perhaps as a sign of good genes and health).

As we have seen before, our individual human body odours are complex and highly individualised, being a mixture of more than three hundred chemicals, all genetically regulated in response to the environment. This diversity in our chemical signals explains why some insects find certain people tastier than others. Future anti-mosquito tactics may involve covering ourselves with minute quantities of other people's (or animals') body odour that are particularly repellent to mosquitoes but undetectable to ourselves.

Travel sick

Since F could remember, he'd always being nauseous or physically sick in cars or boats. As he grew older the problem reduced slightly. It would, however, recur if he tried reading in a car or bus or if there was a slight swell on a boat. His two children aged six and eight were quite different: one was just like him and the other like his wife, who seemed immune to all these problems. When aged 35, F needed a hernia operation. After the procedure he had to spend extra time in hospital due to the effects of the anaesthetic, which left him nauseous and groggy for days.

Travel or motion sickness is strongly genetic and very variable between people. Our twin studies demonstrated that it was 60 per cent heritable. Some people are born sailors, who are never ill, and others so sensitive they feel sick watching *Mutiny on the Bounty*. Even trained astronauts get motion sickness for the first week they are weightless in space. The advantages to those who cope well with movement are obvious: they are able to perform tasks such as fishing or hunting from a boat or moving object. The evidence of man's first use of boats was forty to fifty thousand years ago in the extraordinary Polynesian migrations to Australia, and, closer to Europe, occurred only fifteen thousand years ago in the Mediterranean. The fact that genes for motion sickness have persisted suggests that perhaps the genes for coping with movement just weren't that useful to our ancestors, who didn't get to do much deep-sea fishing, long car journeys or space travel.

The nausea and sickness some people feel after a general anaesthetic is also highly variable, some people having major problems for days and others recovering after a few hours. Our twin study also showed this to be genetic, with the same genes involved that lead to travel sickness, suggesting some common mechanism. It is now believed that motion sickness is caused by an inability to balance the nerve signals that reach the brain from the balance (vestibular) system present in each ear. In susceptible people the abnormal unbalanced signals trigger the brain area that controls nausea and vomiting. Genes and therapies can therefore act at a number of different levels in the system.

Whether nausea, vomiting and motion sickness arose merely due to a change in our recent environment is unclear. They could have had some protective function. In pregnant women, the classical symptom of morning sickness and aversion to rich foods, particularly animal products – meat, fish and eggs – lasts for the first three months and then usually resolves. This could be an ancestral device for avoiding food poisoning in the crucial and early stages of pregnancy when the foetus is most vulnerable. Seasickness may be nature's way of telling you not to eat food while being thrown around in a boat. A simpler explanation is probably more likely – that the genes of many of us have not yet had time to evolve to help us become sailors or read the newspaper on the bus.

Early wrinkles

S was a fit 35-year-old woman who had always been proud of her skin. She had few blemishes, hardly any acne as a teenager and a youthful healthy appearance. As she aged she noticed wrinkle lines on her face appearing much faster than those of her contemporaries, and much more than on her husband – even though he was older. She also noticed blotchy patches of sun damage on exposed areas of her shoulders and chest from many weeks per year of carefree sunbathing. Her job was admittedly stressful and often involved long hours – but was this really to blame for her skin?

The skin is made up of three main layers: an outer epidermal layer, a dermal layer and, underneath, a layer of fat cells. Of these, the dermal layer, consisting of collagen for structure and elastin for flexibility, is crucial. A number of factors affect skin ageing and wrinkles. These include the amount of skin fat, pigment and sun exposure, moisture and thickness. Twin studies we performed in women of a range of ages between thirty and sixty, showed that the thickness of the dermal layer (measured by ultrasound on the face and forearm), decreases slowly with age, but is also highly genetically influenced, identical twins having very similar layers and appearances. Other than genes, we found that cigarettes, menopausal status, sun exposure and age were important. As we age, the outer epidermal layer of skin becomes thinner and drier and the cells replicate more slowly – but it is the lack of support from a thinning or damaged dermal layer underneath it that actually causes the lack of smoothness and creases and wrinkles.

Genes can affect the health of the dermal skin layer in various ways. By altering the amount of natural oils (sebum) secreted by the glands in the skin you can produce oily skin that predisposes you to acne and spots. If not enough sebum is produced this results in dry, brittle skin that looks great when young, but is more prone to damage later in life. This potential reproductive advantage may be one reason why the genes that predispose to wrinkles are so common. Genes may also alter the resistance of the stretchy (elastin) and firm (collagen) components of the dermal layer to damage. Furthermore, dermal damage may also be due to the effects of losing oestrogens, as wrinkling accelerates after the menopause and genes are involved in the sensitivity and response to estrogens. We also found that women with thinner dermal layers had thinner bones (osteoporosis). This relationship may be related to inheriting a certain collagen type, which is also important in the strength and structure of bone.

The darkness of the skin is due to the amount of melanin pigment produced in special cells in the epidermis. This evolved in our ancestors as we lost our hairy covering and has an important protective function against the harmful effects of sunlight, which releases destructive enzymes, affecting our cells and disrupting our DNA. For this reason, fair-skinned people are more prone to 'sun damage', which alters and disorganises the dermal layer, which can't fill out a thinning epidermis. Another indirect genetic mechanism is the amount of fat below the skin. We've seen how the total amount of fat and its distribution on our bodies are highly genetic. Those people with chubby faces and more fat below the skin wrinkle much less than expected – although we all lose these fat cells as we age.

Smoking is the biggest preventable cause of wrinkling. In our study we found smokers had up to 40 per cent thinner dermal layers than their nonsmoking twin sisters, with an even greater effect in heavy smokers. This fact was used to good effect in a governmental antismoking poster campaign in the UK. Acne, as we saw earlier, is highly genetic and is associated with a moister skin, which protects later against the effects of ageing. There appears to be a genetic trade-off between early good looks and greater chance of later wrinkles, or having greasy skin early in life, which will age more slowly. Your parents' skin is a good guide to your future looks and the consequences to your skin of excessive sunlight and smoking. Companies are now developing anti-ageing cosmetics that will be tailored to your precise genetic makeup. The skin is one site where, in the future, gene therapy using specially absorbed creams or injectable younger cells is showing signs of promise – and could put plastic surgeons out of business.

Colds, sniffles and HIV

A suffered continuously from minor illnesses and colds, which, although never serious, left him tired and depressed. He contracted six or seven colds a year, in both summer and winter, and they took around ten days to go away. The infections sometimes spread to his chest or gave him problems with his sinuses. He took large quantities of vitamin C tablets, echinacea and other herbal remedies – but, whether he took them preventively or at the first sign of trouble, they never seemed to do him any good. His girlfriend, in contrast, never seemed to catch his viruses and would complain of a minor cold less than once a year. She thought he exaggerated his symptoms to get sympathy.

The burden of the common cold on humans is enormous. In the course of a year, individuals in the United States suffer 1 billion colds. In the UK alone, in any one day, 930,000 people will have a cold with 279,000 absent from work, suffering from cold symptoms. This means that over a full year more than 100 million working days are lost, costing UK industry at least £2.8 billion per year.

Children have about six to ten colds a year. In families with children in school, the number of colds per child can be as high as twelve a year. Adults average about two to four colds a year, although the range varies widely. Women, especially those aged twenty to thirty years, have more colds than men, possibly because of their closer contact with children. On average, individuals older than sixty have fewer than one cold a year. Also on average, one person out of sixty will currently be suffering from a cold.

More than two hundred different viruses are known to cause the symptoms of the common cold. Rhinoviruses (from the Greek *rhin*, meaning 'nose') cause a third of all adult colds, and are most active in spring and summer. More than 110 distinct rhinovirus types have been identified. These agents grow best at temperatures of 33°C, the temperature of the lining of the human nose. Coronaviruses are believed to cause a large percentage of all adult colds in the winter and early spring. Viruses responsible for other, more severe illnesses, such as adeno-viruses and coxsackie viruses, cause approximately 10–15 per cent of adult colds. The exact causes of more than a third of all colds, presumed to be viral, remain unidentified. The existence of hundreds of different forms of viruses explains why we never become immune to their effects – in contrast to other viruses, such as measles.

These tiny viruses cause cold symptoms by a series of steps. First, several hundred combine in a sneeze or mucus droplet from a previous victim. Then, via the air or hand, they enter the new nose, sticking on to and then invading the cells lining it, breaking into them and forcing them to start making more copies of the virus. This alerts the body to defend itself with its own white cells – a subset called T cells, which stir up a variety of chemicals (cytokines) that try to destroy the virus. Unfortunately, while the tiny virus itself is harmless, its presence leads to an inflammatory action that causes nasal congestion, mild fever, and a runny nose as side effects.

Careful studies by common-cold units in various countries have shown no effect of climate or temperature, chills or vitamin C supplements on whether you develop a cold. Although environment, overcrowding and having young children play a part, there is likely to be a strong role played by the genes that control the response of the body's defence mechanism – white blood cells and the cytokines they produce. Our twin study of more than two thousand adults shows a clear genetic effect: the number of colds per year in identical twins was more similar than in nonidentical twins (twin pairs in both groups no longer lived together).

A small but detailed study of adults in Germany showed that subjects had different forms of genes that influenced both the amount of chemicals (cytokines IL-2, IL-6) released by white (T) cells and how easily the viruses stick to the cell (ICAM-1). Individuals with particular forms of the genes that influence these chemicals (IL-1, IL-2, ICAM-1) had two or three times the expected numbers of colds per year. Tests show that our noses often harbour viruses without our noticing. The lucky people who rarely suffer from colds may well become infected just as often, but don't respond to them as vigorously. The advantages of robustly fighting viral infections are obvious – but why should some people have genes that make them more susceptible? One explanation could be that their defences are more specialised against life-threatening infections or common infiltrations by parasites (such as worms). Having variety in the way people are susceptible and respond to infection also stopped certain diseases spreading and tribes being totally wiped out in one go. When families get colds they rarely all succumb at the same time. Another possibility is that in evolutionary terms there is a trade-off between the amount of energy our genes invest in fighting off infection, and reproducing. We will discuss this concept again later.

We believe that the influenza (flu) virus began to affect humans only when we started living closer to each other and domesticating animals around eight to ten thousand years ago. It mutated from the variety that affected pigs and ducks. Flu is closely related to cold viruses but is more toxic and dangerous, suppressing the body's immune response and multiplying much faster. In recent years it has got milder, but is still one of the major killer diseases of the West. It wiped out 21 million people in Europe just after World War One – twice as many as died fighting in it. The genes of the flu virus are continuously mutating, and replicate rapidly, so we have no specific defences against the next strain. Every year the scientists making the vaccines struggle to keep up with the next wave of potential viral epidemics, which for some reason originate from southern China. The recent SARS (severe acute respiratory syndrome) virus and epidemic originated from the same area. It has been found by DNA matching to be a par-ticularly nasty form of the common cold virus (coronavirus) – with no direct cure. It may have originated in exotic animals and spread to man when they appeared on Chinese menus.

AIDS, which is due to the HIV virus, is a good example of genetic mechanisms for attack and defence. HIV mutated from another form that affected monkeys in Africa – affecting humans only since around 1959. Genes in humans control the levels of a subtype of circulating defensive white T cells called CD4 and CD8 cells (also called helper and suppressor cells). Exposure to the AIDS virus results, in most people, in a depletion in the number of CD4 helper cells, which dampens down the immune response, allows the virus to replicate and flourish and causes the human host to die from a variety of other diseases to which they are now very susceptible.

Initially it was thought that everybody who got the AIDS virus succumbed to it. In the 1980s the onset of illness and death was often very rapid. Surveys in many countries have identified groups of people who definitely have the virus – yet have not developed any of the clinical manifestations. Their immune system appears to be containing the virus, without destroying it. Tests showed that, well before these people were infected, they had higher CD4 counts than those who later developed the clinical symptoms. The reason they had higher protective CD4 counts lies in their genes. Another gene that affects susceptibility to HIV infection and then AIDS is CCR5. CCR5 affects the way in which the virus binds to its receptor. People with a deleted form of the gene, called Delta 32, appear to be protected against transmission. Because Delta 32 is rare in Africans and Asians, it may explain the increased risks in those areas. Progress is being made against the AIDS virus, but it is a formidable adversary, with one of the fasted genetic mutation rates seen – evolving with frightening speed. If it follows the pattern of other viruses, it will evolve to be less lethal to the host, allowing more time for it to spread to other humans.

Viruses are one of the simplest stripped-down forms of life, being just DNA and a lining, a twentieth the size of the cells of our body. All our genomes contain large numbers of inserted viral genes, acting like hitchhikers, that manage to get humans to replicate their genes for them without all the bother of infecting us each time. They are probably our very distant ancestors, who have managed to survive so well by the policy of 'keeping things simple' and constantly changing. This has worked for over 5 billion years and as a species they are likely to long outlive humans – but we shouldn't envy them their lifestyle.

It's on the tip of my tongue

V had always prided herself on her ability to remember things. At school she had what she considered a 'photographic memory' that allowed her to memorise pages of books by visualising the text and position of sentences and words. By the time she reached 45, her skill was starting to fail her. Names of people were a particular problem, which could be socially embarrassing. She was also often forgetting where she had left the keys – and sometimes the car. She wondered whether these changes were an inevitable part of ageing, lack of mental stimulation or even genetic, as her mother's memory was now very poor.

Memory is but one facet of our overall cognitive abilities or 'intelligence'. There is wide variation between people in their ability to recall information. Memory is a series of chemical and nerve-conducted impulses in the brain, and comes in several forms. Short-term working memory lasts a few seconds, such as when we memorise a phone number from the directory before making a call, and then evaporates. Long-term memory is different and involves a series of complex chemical reactions in a small part of the brain called the hippocampus (Greek for 'sea horse'). Painful or traumatic memories are stored in the fear centre of the brain, the amygdala. Losing the keys involves the hippocampus, and is a failure of the brain connections used in spatial learning to produce a solid mental map of where the keys or car were left. For many people the frustrating experience of being unable to recall names is a common occurrence. Research has shown that this involves different areas of the brain that are more difficult to access. This is why, although you can't remember names spontaneously, you can recognise them easily if prompted – and can usually describe in great detail the person, film or whatever you are thinking about.

Twin studies covering a range of ages from five to eighty years have shown the importance of genes on visual and verbal memory, although, contrary to what we might expect, this effect actually strengthens with age. It appears that as we get older, our environment has less and less influence. In elderly Swedish twins over the age of eighty, memory was over 50 per cent heritable, considerably more than for the infants, showing the importance of early training and the ability of young people to improve their memories. In our own twin studies we showed that skills involved in specific computerised memory tests, such as recall of visual images, were also heritable, as were simple traits such as often losing keys.

Methods of memorising are also genetically influenced. Photographic or 'eidetic' (Greek for 'image') memory is the ability to form complete visual images in your head. One study examined school pupils by showing them a picture from *Alice in Wonderland* and assessing their recall of details, such as the number and colour of stripes on an animal. Four per cent had this special skill. One famous memory expert (mnemonist), a Russian called Shereschevsky, had unlimited powers of visual recall. He could recall nonsense words or formulae given to him briefly fifteen years earlier and remember exactly what the interviewer was wearing at the time. While he used some memory-training techniques, most of his skills were innate. Indeed, his brain was so full of detail that he was often distracted by his own memories and had difficulty thinking abstractly. He was unable to hold down mundane regular jobs and died an impoverished entertainer. Photographic memories, therefore, don't always provide the advantages in life one might expect.

So, if a large part of memory is genetic, is there anything you can do about it? There is some evidence that old people who keep working, reading, performing arithmetic and solving crossword puzzles late in life perform better. We don't yet know with certainty the genes involved, but ApoE (the dementia gene) is one candidate, as well as more general ageing genes. In mice and other laboratory animals, the importance of genes that control a key substance called CREB in the hippocampus has been demonstrated. Blocking this protein causes animals to lose their long-term and spatial memory and injecting the substance miraculously restores it. Several companies are now pursuing this as a treatment in humans – allowing us to contemplate the potential and pitfalls of so called 'smart' drugs, which we might be popping before exams, parking the car or even before cocktail parties so we can remember people's names.

Sleepless nights

F and her husband K were poles apart when it came to sleeping. F always fell asleep instantly, then slept like a log for an average of over eight hours per night. She had tried shift work for a while – but couldn't cope with the fatigue and also didn't adjust well to jet lag. K had trouble getting off to sleep, but could usually survive on six hours per night. K started to sleep badly when work became more stressful, and he needed to wake up earlier. He used sleeping tablets when he was stressed at work and tried avoiding coffee completely. They used to quarrel over bedtimes, as she was always asleep and snoring when he got into bed.

We sleep for a third of our lives. We know it helps the brain to recover and aids memory – yet the detailed biological reasons for sleep are only now becoming apparent. Lack of sleep is a major cause of accidents and psychological and physical problems. There is large variation between individuals in sleep patterns – in terms of hours needed, sleep disturbance and deepness of sleep. Some of this is influenced by age: we need less sleep as we age. It was commonly believed, until recently, that training and culture were mainly responsible. However, it appears to be an in-built trait. Twin studies from different countries, including ours in the UK, have found that sleep patterns are around 60 per cent heritable. Lifestyle variables such as coffee consumption, alcohol, cigarettes and culture accounted for only a small fraction of differences between people.

It is now realised that the brain tightly controls the amount, depth and duration of sleep. The area responsible is again the busy hypothalamus, which contains specialised regulatory areas that overlap with the control of appetite, body temperature, body stimulation and mood. This explains why sleep deprivation increases appetite and alters temperature and mood. A tiny area of the hypothalamus (called the SCN) contains cells that produce our 24-hour clock, or circadian rhythm. The body clock appears important in determining which activities are best performed at certain times. For example, intellectual activities are best in the mornings, and physical activity is best in the afternoon, when most world records are set. These SCN cells are influenced by a number of hormones, such as melatonin and cortisol. Natural long-sleepers like F have increased levels of melatonin in the blood and a delay in the natural early morning dip in temperature and rise in cortisol, compared with short-sleepers like K. This tells us that these patterns and hormones are in-built and difficult to change.

Other sleep-related conditions are also important and shown in our studies to be equally genetically influenced. Heavy snoring is one that is often linked to obesity – with a suggestion from our twin studies that some of the genes overlap. Although the subject of jokes, snoring can often lead to serious health problems such as early heart and lung disease and premature death. The ability to recall dreams is another. Related symptoms like 'restless legs' or twitching of the legs at night are also strongly genetic, and may be a normal body response to keep the muscles working, but one that is exaggerated in some individuals.

Genes for a few rare sleeping disorders have recently been discovered that give new clues to the process of sleep itself. One of these is narcolepsy, in which the unfortunate sufferer classically can suddenly fall into a deep slumber day or night, with no warning. Relatives of affected people have a thirtyfold increased risk. The gene for this is the hypocretin gene on chromosome 17. This regulates the amount of the hormone of the same name in the hypothalamus. Sufferers have very low levels. Now genetic testing is possible, it looks like variations of the disease are far commoner than suspected. Current estimates suggests it affects one in 2,500 people and occurs in a milder form, causing occasional episodes of excessive daytime drowsiness – a symptom we can all relate to.

Sleepwalking is another rare trait that has a genetic component. It affects only 1–4 per cent of adults, but one in five children. One recent study has shown that an immune gene, HLA DR4, may be involved. Narcolepsy also has a strong link to certain immune genes (HLA DR4 and DQB1), suggesting a genetic link between sleeping and our protective immune systems. This may not be so crazy, because sleep does alter our chemical responses to infection (cytokines) and these same substances alter sleep patterns in laboratory animals and may have a similar effect on us.

Other genes have been found that upset the circadian patterns of certain rare families. A single mutation in one gene, the 'clock gene', caused a very unusual family from Utah (and probably antisocial neighbours) to go to bed at 7 p.m. and wake at 2 a.m. – not a great family to live next to. A number of other genes, such as the period 3 gene, have also been shown to have similar effects, and, in a study of five hundred people visiting the Science Museum in London, this gene was found to associate with their being 'morning' or 'evening' people. These sleep genes should hopefully provide future therapies for jet-lag, shift work and possibly bad sleepers.

It is not clear why we vary so much in sleep patterns. Our personalities do influence our patterns to some extent, but environmental factors are also important. The best sleepers are married men with no kids – they sleep well in the night and day. The beneficial effects of marriage or living with a partner for men are, however, counteracted by children. Education is also a factor and is associated with shorter sleep duration in both sexes. Perhaps there were evolutionary survival advantages of having a variety of slumberers in any tribe, so they weren't all asleep while being attacked. It may also be a balance between productivity and survival advantages of short-sleepers against the restorative effects on your brain and immune system of long-sleepers.

The main lesson is that you can't dramatically alter your natural sleeping habits merely by alarm clocks and timetables. Eventually your body's needs – based on ancient genetic pro-gramming – will have to be met. There are probably advantages and disadvantages in life to all kinds of sleepers. Some jobs select for people who can survive on little sleep, such as emergency work or high-profile politics. Margaret Thatcher reportedly slept for only four hours per night, which may be very effective for a head of state but not ideal for domestic bliss.

Medicines don't agree with me

H had never been able to take tablets or medicines without problems. Sometimes she came out in a rash; more often she just felt sick and nauseous. Her doctor put her on a tiny dose of an antidepressant that she took at night. The next morning she felt like a zombie and was drowsy all day and had to stop taking them. She was always a bit anxious and her doctors attributed her problems with medications to this, more than any rare side effect effect of the drugs she was taking.

As with the response to alcohol, the way we react to all drugs and medicines varies enormously between individuals and is predominantly genetic. One variable easily measured is how long the drug stays in your body (called the half-life). In twins used as guinea pigs, heritabilities of around 60–70 per cent have been shown for most medications so far studied – including aspirin, antidepressants, sleeping pills and anti-epileptics. Thirty-six-fold differences in half-lives have been found between individuals as well as differences in the speed of action, how long it works for, the best dose needed and the side effects. Some people inactivate (metabolise) drugs fast and others slowly. There are many known drug-metabolism genes that make the enzymes that inactivate the drug and eliminate it from the body. The best known are the cytochrome p450 family (known as CYP genes), which are found in the liver. Differences in these genes, of which we all have many, determine many of the varying ways we react to drugs and poisons.

A good example is people who take the blood-thinning drug warfarin for heart or clotting problems. Some people, when given the normal average dose, will get major bleeding problems, because their body can't form clots. These people have a sixfold chance of having a special variant of the gene (called CYP2C9), which doesn't work as well. In a few hospitals they now test people before giving them their first dose of warfarin. The same gene also can lead to overdoses with the drug for epilepsy, phenytoin, and some antidiabetes drugs that lower blood sugar.

The commonly used group of antidepressant drugs (called tricyclics), of which there are around thirty different types on the market (imipramine, amitriptiline etc.), all depend on a process of inactivation by two of the CYP genes (CYP2D6 and CYP2C19). People can be categorised as good, intermediate and poor metabolisers – and progress has been made at recommending different dosages. About one in three patients doesn't respond to epilepsy drugs, and this has recently been found to be due to differences between people in their ABC1 drug transporter genes – a test for this could soon be used to monitor patients. Many psychiatric drugs work by altering levels of brain serotonin via its receptor. Variations in the serotonin receptor gene can alter response to drugs like this, such as the anti-schizophrenia drug clozapine, or the weight-reducing drug fenfluramine. Other enzymes (called TPMTs) are important in metabolising the dangerous but life-saving drugs that treat leukaemia and suppress the immune system. Even small doses in people who are poor metabolisers can sometimes be fatal.

Why do some people prefer aspirin to paracetamol? Why do others feel sick with antibiotics or anti-inflammatories such as brufen? Genes are the likely cause – but we don't yet know which ones. Drug manufacturers have been reluctant to invest money in this area for popular drugs already on the market. From their point of view it is better to get more people to try them, even if a proportion feel sick. Only when the side-effects are serious or life-threatening and the drug may be withdrawn are they interested in providing genetic data.

Another overlooked area that shows wide variation is blood monitoring during therapy. Millions of patients regularly have blood tests to see how they are responding to drugs for diabetes, thyroid or blood thinning. Levels of blood sugar, thyroid hormones or clotting times are measured and the doses of drug adjusted. The fact that all these blood values naturally differ between us quite markedly is often ignored. This means that someone who has naturally high thyroid hormone levels will be told they are taking too much thyroid replacement, even if they complain that they are tired and putting on weight, as (for their individual genetic makeup) they have too little.

Unfortunately, manufacturers and laboratories base their results on 'average' values, although very few of us are exactly average. In the future, drugs and doses used will be custom-designed to our own genetic makeup and we will be tested to exclude high risks of side effects. Until then you need to be aware of the very different and individual ways you may respond to medications, and don't be worried to report to your doctor that a dose suitable for a mouse still makes you sick, or that a dose of sleeping tablet that could knock out a rhino doesn't touch you.

Baldness and macho genes

M had always been proud of his looks and wanted to look after his body. He worked out regularly in the gym and was muscular with a good covering of body hair. His father and grandfather had both gone bald in their late thirties, but he was relieved when his brother, who was five years older, showed no signs of any problems. His mother had gone grey at the age of 28, but so far he saw no sign of any grey hairs on himself. However, when he was thirty he started to lose his hair rapidly around his temples. He read that it was due to overactive male hormones – but didn't know whether this was just something said to make people feel better. His sister wondered whether she was likely to have problems of thinning hair as well.

Baldness of the male pattern or common type is predominantly a genetic disorder. It is very frequent. affecting 50 per cent of European men by the age of forty, but is less frequent in other groups, such as those of recent Asian or African origin. The process of hair loss centres on the hair follicle, which undergoes 'miniaturisation', producing only a tiny hair shaft in the affected area. The total number of hair follicles is only slightly reduced. Rarely, rapid loss of hair can be associated with other medical problems or drug allergies. Genes for very rare causes of baldness that lead to sudden and total hair loss (*alopecia universalis*) have now been identified from a Pakistani family that had a gene on chromosome 8 (called originally the hairless gene). The genes for a type of baldness caused by an immune reaction (*alopecia areata*) have been linked to the immune and HLA system. The many genes underlying the common forms of baldness are as yet unknown – but the main suspects are those involving hormones and steroids.

The genes involved undoubtedly alter the response of our 5 million hair follicles to the male hormone testosterone and its derivatives. What appears to be crucial is the way circulating testosterone is converted by an enzyme (5–AR) in the follicle to its more active form, DHT. The search is on for genes and drugs that control this enzyme. What is curious is that DHT makes some hair follicles grow and others shrink. The process works slowly and follicles shrink gradually over about five years, which is their normal life cycle.

For men, a good guide to future baldness is to look at your father and ask about your grandfathers. If both grandfathers had a reasonable head of hair it is highly unlikely you will go bald, and vice versa. For women, baldness in a milder form is more frequent than you might expect. In our twin survey, one in six (15 per cent) of women complained of thinning hair – often causing considerable distress. Most of these women had a family history of baldness on the male side.

Although some companies offer commercial genetic testing (focusing on hormonal genes) for susceptibility to baldness, these are bogus, as no gene has yet been conclusively identified with hair loss or treatment. The stories that bald men are more virile come partly from the findings that castrated males (eunuchs) hardly ever go bald. The other pointer is that the anti-male-hormone drugs such as finaseride block DHT production and increase hair in males, but unfortunately, due to their effects on the rest of the body, cause impotence. Despite this, there is no proven link between having a shiny scalp and the body's circulating testosterone levels, libido or sexual prowess – although, as consolation, few men would swap baldness for impotence. Modern surveys suggest that women in general find men with hair on top more attractive, but perhaps follicularly shrunken men should seek solace in the fact that many early European women may have preferred bald men with hairier chests and the tradition might return.

Going grey is very common and eventually affects all of us at some time. It signals when the pigment cells in the hair follicle stop working. Contrary to popular opinion, you can't go grey overnight, because the lack of pigment shows only as the hairs grow, and this takes several months. Genes are responsible for the wide variation between people when this happens. It's unclear whether our environment or stress has much effect at all – although there are few proper studies.

Whether genes for early greying are random or have some secret purpose is not clear. One idea is that grey hair has been selected for, particularly by female preference for men, as a sign of wisdom and survival. Veteran male gorillas in charge of large harems of females have abundant grey hair (and so are known as 'silverbacks') and are easily recognised by younger adversaries and willing females. Greying (silverback) film stars such as George Clooney and Richard Gere can have the same effect on female fans. Scientists have been able to add genes that change the melanin and hair colour of mice, so grey hair is potentially preventable, although some males might risk losing their harems.

Toothless

T was always visiting the dentist when in her teens. She had many fillings but, as her diet improved and her tooth brushing became more regular – these problems diminished. Aged thirty, she developed bleeding gums and needed to have two teeth removed. Her dentist informed her that this was now mainly due to gum disease, which may get worse. She was at risk of losing her teeth. She was concerned that her mother and aunt, who were now in their fifties, had already lost all of their teeth. T could not understand how her sister never had any problems – despite having a similar diet and a life-long obsession for Coca-Cola.

Losing your teeth is unfortunately very common. A quarter of people in developed countries have lost all their teeth before the age of 65. The commonest causes of tooth loss are periodontal (gum) disease and caries (tooth decay). Dental caries is one of the most common of all disorders affecting the majority of the population in developed countries. In a recent UK survey of a hundred thousand twelve-year-olds, 38 per cent had some detectable caries, although there were marked regional variations – with Wales and Scotland having the worst rates. It is the principal cause of tooth loss in younger people

Bacteria normally live in our mouths in large quantities. They convert all foods – especially sugar and starch (from carbohydrates such as potatoes) – into acids. Bacteria, acid, food debris and saliva combine in the mouth to form a sticky substance called plaque, which adheres to the teeth. It is most prominent on the grooved chewing surfaces of back molars, close to the gum line on all teeth, and at the edges of fillings. Plaque and calculus (like lime scale) irritate the gums, causing swelling and loosening of the teeth. The acids in plaque dissolve the enamel surface of the tooth and create holes in the tooth (cavities).

Gum disease is primarily caused by a build-up of plaque between the tooth and the gum, and also involves bacteria. In a recent US survey, about one in three Americans aged over 65 was toothless – with rates varying markedly between states. Hawaii had the lowest with only 14 per cent and West Virginia was the worst state with nearly one in two people affected. Surprisingly, these rates are actually 15 per cent better than in 1985. These differences between regions implicate both socioeconomic and genetic factors.

The effect of genes on dental problems has been largely overlooked by dentists, who believe in the importance of environmental factors. This is due to changes to our diets and brushing habits over the last fifty years. These have led first to steady increases in caries and periodontal disease, and then to spectacular recent falls in caries. There has been a 20 per cent reduction between 1978 and 1998 in adults in the UK and even greater reductions in children. Now only 13 per cent of UK adults have no teeth, compared with 30 per cent in 1978. This is mainly due to the widespread fluoridation of toothpaste and drinking water. Working out exactly how genes affect us is more difficult, because no clear genetic influence on levels of plaque-forming bacteria in our mouths has been seen. This suggests that it's not the amount or type of bacteria in our mouths that is crucial, but our responses to it, and to our sugary diet. An interesting natural experiment happened in 1961 on the desolate volcanic island of Tristan da Cuhna, where there was no sugar. The volcano erupted and the population, descended from British servicemen, were evacuated to England. They were extensively examined and found to have no evidence whatsoever of tooth decay. Within twelve months of the sugary diet back in their motherland, they all had cavities.

Because of the big dietary differences between generations, family studies are of little help in separating out the effects of diet, oral hygiene and genetics. Twin studies of dental problems have now been performed in rare sets of pairs reared apart. The researchers found that, despite the fact that subjects had been reared in different environments, receiving different diets and different professional dental care, the identical twins had very similar teeth. There was a considerable genetic effect (heritability of 45–67 per cent) on the number of teeth remaining, the number of teeth repaired, and the number of cavities. A classical twin study in 117 pairs of US adult twins reared together has now confirmed that gum disease (periodontitis) was 50 per cent heritable and unaffected by differences in dental hygiene. Thus, even if the diet contains too much sugar, brushing habits are poor, and smoking is a known risk factor for gum disease, there is still a strong genetic predisposition to dental problems. Mouth ulcers, which are common in children and some adults, are blamed on vitamin deficiency, stress or mysterious infections – but twin studies have also shown them to be highly heritable.

Genes that now predispose to tooth decay may have had some protective advantages, such as preventing major mouth infections. Our Pleistocene ancestors were never exposed to frequent sugary or starchy foods, so never developed caries, and protection against gum disease was probably effective until middle age and after reproduction. Unlike us, they didn't live long enough to find out if their teeth would fall out, but, being regular eaters of tough raw vegetables and uncooked meat, they wouldn't have lived much longer if they did fall out.

PMT genes

Since her periods started, S always had major mood swings, which were greatest for her around days 23–25 of her menstrual cycle. Her three children and husband learned to predict her mood changes and knew it was wise to postpone a difficult discussion or argument for a few days. Over those two to three days she felt bloated and unattractive. She was sometimes prone to short bursts of tearfulness and occasionally threw plates in her husband's direction if he annoyed her. As she progressed towards her menopause at age fifty, her symptoms lessened and eventually to her relief, disappeared.

We have already seen how the timing of the first period (menses or menarche) is under the control of genes and triggered by body size. The regulation of the periods into fixed cycles is also strictly regulated. Three-quarters of all women complain at some time in their lives of premenstrual symptoms, with the most common complaints being weight gain and swelling, breast tenderness and mood changes. A more extreme subtype is called PMD (or premenstrual dysphoria), which affects about 5 per cent of women and is characterised by more extreme mood changes and irritability. Women generally seek help in their thirties.

The genetics of PMT has been studied in a large Australian twin study. More than seven hundred pairs were asked whether they suffered any of seventeen symptoms of PMT as well as other psychological traits. One in four women met the official definition of PMT. A clear genetic influence, explaining 44 per cent of symptoms, was found. Although there was some overlap between genes for anxiety and neuroticism and

PMT, there still appear to be specific genes that influence a woman's response to cyclical changes in hormones

The mechanisms of PMT are poorly understood, but are believed to involve a lack of balance between the fluctuations in the sex hormones and the brain transmitters, such as serotonin, which act on the hypothalamus to affect weight and emotion. Treatments for the more severe forms are Prozac-like drugs that alter serotonin levels and can be used just for the few 'critical days'. The oral contraceptive pill is often used as an alternative for milder cases and works to smooth out the hormonal fluctuations over the cycle. Chemically created pheromones (as nasal sprays), similar to the natural agents produced by males, are now being used as potential treatments for PMT, as they have been shown to have a calming influence on sufferers.

For most women not taking synthetic hormones, oestrogens produced during the first 21 days create a feeling of wellbeing and, when testosterone is also produced, increases libido around mid-cycle. After day 21 the hormones decline quickly and in susceptible women, PMT and mild depression can occur, leading to increased marital friction, increased rates of car accidents and sometimes even criminal acts. Interestingly, PMT has sometimes been successfully used as a reason to acquit women of criminal charges – in contrast to the legal view of the effects of testosterone on men.

Women, when living closely together in groups, have evolved a system of synchronising the timing of their periods. A series of recent experiments found that women subconsciously do this by recognising certain pheromones from other females' armpits and sweat glands, and within a few months around two-thirds of women's cycles will be closely coordinated. This phenomenon doesn't require close or intimate contact and is seen in women living in the same house in different bedrooms.

Pleistocene woman rarely experienced regular periods, as they were often pregnant or breast-feeding for several years. Modern woman, in contrast, can 'look forward' to around five hundred periods in a lifetime and so feels the side effects more often. We can only speculate why this cyclical and co-ordinated pattern evolved. One controversial idea we discussed previously is that it evolved as a way of testing the patience and supporting instincts of males. As menstrual synchrony is seen in other animals – rodents, hamsters and some apes – it may just be a piece of evolutionary baggage that was useful. It may have enabled females (by being fertile at the same time) to prevent themselves being impregnated by the same mates. Another theory is that it may have evolved to avoid jealous or violent confrontations. When a male returned from hunting with sex in mind he would find his chosen partner uninterested and 'indisposed' – but, unfortunately for him, would have no temptation elsewhere.

Flying risks

K was a 48-year-old commercial lawyer who was often flying around the world on business. He thought no more of the potential risks than taking a train. On one flight from London to San Francisco he was relaxing in business class after a heavy week of work with a big gin and tonic and fell asleep. When he awoke four hours later he had an uncomfortable feeling in his left calf, which was slightly swollen and tender. The pain grew steadily worse over the next hour. He also started coughing and noticed this too was painful when he took deep breaths. He alerted the cabin staff and was given oxygen as he left the plane and sent straight to hospital. He was later diagnosed as having a clot in the veins of his leg (a DVT or deep-vein thrombosis) and his lungs (pulmonary embolus). He needed blood-thinning treatment (with war-farin) for several months and didn't fly again for a year.

Blood clotting in veins is a major worldwide health problem. Clots leading to blockage of the lungs (pulmonary embolism) caused the deaths of 175,000 Americans in 2002, although most of these were unrelated to air travel. It is also the leading cause of maternal death in childbirth. Previously, the major causes of clots were thought to be drugs and hormones, ill health and prolonged immobilisation. The perception and profile of the disease changed in 2000, when 28-year-old Emma Christofferson, a fit, sporty girl who ran marathons, died of a clot in her lung after flying from Sydney to London. Pulmonary embolism was renamed 'economy-class syndrome' due to the cramped conditions and lack of legroom – although there is no clear evidence that it happens more in economy class than in business class.

It appears that flying does increase your chances of a clot, but not much more than long car journeys without breaks. Studies have, however, shown no clear link between alterations in aircraft cabin pressure and DVT risk. There is still considerable controversy over the amount of risk involved, as the average person who stays on the ground has a one-in-ten-thousand risk per year of a DVT. Around 1.7 million Americans fly every day yet encounter only just over one major medical problem per billion miles flown. Despite this, many people take their own precautions, regardless of medical advice, such as ingesting aspirin to thin the blood or wearing uncomfortable surgical stockings or avoiding flying altogether.

Most people wouldn't guess that genes are the predominant risk factor: approximately 10 per cent of people of northern European origin have a genetic profile associated with a sevenfold increased risk of the disease. If your genes increase your risk sevenfold and flying increases it theoretically another threefold, the risks might get to as much as one in five hundred per year – still not massive, but worth thinking about.

Of the major genes so far known to be involved, the most important is called Factor V Leiden. This gene codes for a protein (of the same name) that is essential for controlling whether you bleed to death or prevent it by forming a blood clot. The balance is quite crucial. The gene regulates the speed and efficiency of the clotting process, which is highly heritable (60–70 per cent) and is present in about 3–8 per cent of Europeans and in as many as one in four people with clots. As always, there are pros and cons of having these genes. People who clot more easily are probably protected from bleeding excessively from injuries. These genes are more frequent in people in northern Europe than the south – differences that may have evolved as mutations to prevent death due to bleeding in childbirth, which was a frequent cause of death in our ancestors.

Being a good lawyer, K, the man in our case study, looked into suing the airline company, who were already dealing with thousands of claims. Although airlines have not yet lost a lawsuit (the family of Emma Christofferson, who died from DVT, were unsuccessful in suing the airline) they now issue standard warnings on all flights and show stretching and anticlot-exercise videos. K may have to wait a while for any compensation, and, if it's clever, the airline will order genetic tests on him. Some commercial screening tests for clotting risk are now being offered based on a panel of six genetic markers. If proven clinically useful, these tests could be used for targeting people in real need of preventive therapy. These people might include frequent travellers like K, those with a family history of clots and women taking the contraceptive pill or HRT, who have a threefold increase in risk. If you are in the majority group, and don't have the high-risk genes, you can relax and enjoy the comfort of your 32 inches of legroom in economy – knowing your risks of having a clot (or sleeping) are very small.

8

Genes, Diseases and Getting Older

We all develop diseases of one sort or another, and we all share one eventual outcome: death. Overall, one in three people will die of cancer, and over half of heart disease. But, as with every recent generation, we all are now living longer than our parents. Death is now less of an issue than quality of life, and other diseases that cause disability and pain become more relevant to us. Many of these have strong genetic components, allowing us the chance to assess and reduce our personal risks by modifying our habits and environments.

Will I get my parents' cancer?

D was a reasonably healthy forty-year-old with two children of school age. She had never had any major health concerns. One day she learned her mother, aged 65, had, quite out of the blue, been diagnosed as having a rare form of cancer of the womb. Within two weeks, her previously fit mother was dead. After the initial grief and shock had passed, D was suddenly anxious about getting the disease herself and passing it on to her daughters. She had heard that cancer was usually hereditary.

Although cancer is now known to be a genetic disease, it is not, unlike other conditions we have discussed, usually inherited in the conventional sense. When a body develops cancer it involves a mutation of certain genes. This is something that is not directly heritable, as it occurs in the cells that are not passed on to the next generation although the tendency for cells to mutate can be. In comparison with most other common diseases, the influence of inherited genes on cancer is much smaller.

All cells in the body are programmed to copy themselves continuously and repair the losses due to ageing and wear and tear. There are genes that keep this process going (growth-factor genes) and other control genes that stop it (tumour-suppressor genes). These genes are collectively called oncogenes and alterations in them cause cancer. Some cancers are triggered by oncogenes passed to us by viruses. An example is cancer of the cervix (neck of the womb). The cancer occurs only if at least two steps occur. First, you get infected with a virus containing the special oncogene, and, second, your body's natural anticancer defences (tumour-suppressor genes) are overcome because they are not working effectively. You can inherit different kinds of tumour-suppressor genes from your parents, and some faulty ones (like one called p53) can lead to increased risks of cancer.

Until recently the overall genetic contribution to cancers in the population was unknown. A massive twin study of 45,000 Scandinavian twins from the registers of several countries found in 2000 that overall heritability for most cancers was around 30 per cent – about half the figure for most other common diseases. Within these summary figures there were important variations for the different types and sites of cancer. The heritabilities ranged from that for the prostate (42 per cent), to intestines (35 per cent), to bladder (31 per cent), to stomach (28 per cent), to lung and breast (27 per cent) to ovary (22 per cent) and finally to cervical cancer, cancer of the womb and common kidney cancers, for which there is no detectable heritability.

Just because a cancer is not inherited, it doesn't mean your risk is zero, as you may be sharing certain lifestyle or dietary factors (such as smoking). Yet the risks of having the same cancer for first-degree relatives (if your relative has a common cancer) are still only about 3–5 per cent on average. The risk of disease in first-degree relatives is increased only about a third (33 per cent) rather than threefold for most common diseases, such as arthritis, diabetes, heart disease and osteoporosis. In D's case, cancer of the womb has a nearly zero heritability, so her odds of developing the same cancer as her mother are not increased substantially.

There are a few rare cancers that do run in families (usually called 'family cancer syndromes') which are clearly genetic with much higher risks. In general these stand out as having a distinct pattern. They usually affect people at younger ages, often before the age of fifty. They usually affect several family members in the same generation, and can predispose people to more than one type of cancer. For these rarer types of disease, a detailed family history is essential and screening and professional counselling are advised.

In summary, cancer is common but genetic factors are not usually strong compared with other diseases. There are currently no clinically useful tests for genetic predisposition to cancer. However, gene tests are being developed that will in the future be able to detect early mutations in saliva, urine and faeces that will give early warnings. One of these is the APC gene, which is abnormally short in early colon cancer and can be tested in your stools. You should inform yourself about the disease and whether anyone else in your family ever had it. For rare cancers don't worry, and for commoner ones, reduce your own risks by assessing your own environment and lifestyle – smoking, for example, is a risk factor for nearly all cancers, whatever your genetic risk.

Breast-cancer risk

R was a happily married orthodox Jewish mother aged 28 with three children. She was approached by her family about getting screened for breast cancer as two aunts and a grandmother had developed it in their thirties and forties. A Jewish genetic counselling service told her of the possible risks of disease and what choices were available if she took the test. She was not keen on some of the options if the test was positive, such as having both breasts removed, but was told there were drugs (such as tamoxifen) that could reduce her risk as well. She decided eventually to take the BRCA gene test. One week later, after considerable anxiety, the tests came back – they were negative.

Breast cancer is one of the commonest tumours in developed countries, affecting one in ten women over a lifetime. Only 1–3 per cent of breast cancer is of the familial type as in the scenario, occurring at a younger age than normal and affecting several family members. Two genes have been identified on chromosomes 13 and 17, the BRCA1 and BRCA2 genes, which often show special mutations in Ashkenazi Jews – perhaps due to lack of recent racial mixing. The proteins normally produced by these genes protect genes from DNA damage due to natural radiation and other toxic factors. When the genes show a mutation and are defective, the DNA cannot be repaired naturally and cancer often results.

Having these mutated BRCA genes confers more than an 80 per cent chance that the woman will get breast cancer and a 40 per cent chance of her developing a related ovarian cancer at some time. Genetic tests are now routinely used but are very expensive in the USA, where they are controlled by a commercial genetics company. In tests on UK women with cancer aged under fifty, around 3 per cent had these mutations (and fewer than 1 per cent if over fifty) – much lower than in Ashkenazi groups. However, if women with cancer had one family member who also developed the disease before sixty, the proportion rose to 11 per cent, and 45 per cent for those with two first-degree relatives. Unfortunately, these tests are useless for the remaining 97 per cent of common forms of breast cancer, for which no reliable genetic tests exist at present.

There is a genetic influence to the remaining more common forms of breast cancer, but its influence is much less than for the BRCA gene types, with heritability estimates of only 27 per cent with no single genes being involved. Other important risk factors that increase the chances of breast cancer are also partly under genetic influence. These include the age at which your periods start and finish: the longer your reproductive period, the greater the risk. Another genetic factor is the density or thickness of breast tissue. A twin study showed this to be highly genetic, and the denser the breasts the more the risk. Women who have high levels of natural oestrogens are another risk group, although they are protected against the effects of osteoporosis (brittle bones). One of the strongest risk factors is increasing age at the time of first childbirth. For our ancestors, this occurred usually around the age of sixteen compared with the current averages in developed countries, which approach thirty years of age. This probably gave our ancestors less disease, and might partly explain the increasing scale of the problem.

Moles, sun and skin

G was worried when his mum told him and his sister, E, that she had been recently diagnosed as having a type of skin cancer called malignant melanoma. Luckily, it was detected early, before it had spread, and was successfully removed. She warned them both never to go in the sun, as they ran the same risks as she did. They were both concerned by this, as they were sporty and loved the outdoor life. Every time they got slightly sunburned or saw another freckle appear on holiday, they became anxious. Were they right to be worried?

Skin cancer is very common – affecting around 50,000 people in the UK and 250,000 in the USA annually. Ninety per cent of these cancers are not dangerous, as they don't invade other tissues (they are called benign squamous cell and basal cell carcinomas). The other 10 per cent are dangerous tumours called melanomas, which occur in about five thousand people in the UK per year and are fatal in one in four cases. Most people believe that sun exposure is the only important factor. However, we now know that the genes that determine an individual's skin type are crucial. Black Africans, despite endless sun exposure, virtually never get melanoma. Rarely, there are exceptions, like that of the reggae singer Bob Marley, who died from a lesion on the white unprotected soles of his feet. White-skinned people with dark or olive skin that tans easily without burning also rarely get it.

The highest rates in the world occur in pale-skinned people who recently migrated to hot climates, such as Scots and Irish who, two hundred years ago, moved to Australia and South Africa where rates are now five times higher than in the UK. The average risk to a white person in the UK or USA is less than one in two hundred over a lifetime. Your risks increase fivefold if you have a first-degree relative with it, fourfold if you have red hair, and fivefold if you have lots of moles (more than a hundred) on your body. It follows that if you have red hair, pale skin, lots of moles and a family history, you have up to a 50 per cent lifetime risk and need to reduce your sun exposure and to have your moles checked often.

The commoner types of skin cancer, although not lethal, often need surgery to remove them. Dermatologists recognise that people with pale, dry, sun-damaged, wrinkled skin are likely to get squamous cell cancers and those without wrinkles and more moist oily skin, basal cell tumours. All these risk factors are due to genes that influence the pigment in the skin. So why do we have such different skin types?

Our ancestors came from Africa, where it made sense to have skin that didn't burn. As man lost his apelike hair in an attempt to say cool, his skin beneath was probably initially pink and vulnerable. Skin slowly evolved to be black and protective against the damaging effects of the sun, in particular the harmful effect this has on reducing levels of folate in the body (an important vitamin, vital for reproduction and growth). Throughout the world men tend to have slightly darker skin than women – perhaps due to risks of greater exposure. As man later moved to Asia and Europe, then slowly northwards and westwards, his skin became paler. The pale skin evolved to deal with a new problem, the lack of sunshine, which in Africa generated helpful doses of vitamin D – essential for bones and muscles. In the weak sunshine of northern climes, the genes controlling skin pigment mutated to allow better absorption, which prevented diseases due to vitamin deficiency such as rickets. Pale skin absorbs vitamin D five times better than dark skin. Skin colour is therefore a careful evolutionary balance between protection against the sun and folate reduction, and allowing enough of the sun's rays to produce vitamin D.

Hair-colour genes often go with skin colour and the red-haired gene has now been identified (MC1R). This genetic mutation is believed to be a strong marker of Viking ancestry and occurred when Scandinavians settled in small groups derived from families with these mutations. Although we know the Vikings travelled extensively and got to places such as the Americas, these genes (although not the red hair) are now popping up in unusual places and peoples – such as the coast of Africa and the Middle East – suggesting they spread their genes even further than we thought.

Genes for blonde and red hair are known as the recessive type. This means that, as we inherit a pair of chromosomes, if the gene on the other chromosome codes for brown or black hair, the blonde hair won't be expressed. As Scandinavians and other northern Europeans are now travelling more and meeting suitable dark-haired mates, blondes and redheads will slowly become extinct. In the same way, the dark-skin genes of African-Americans are slowly blending with mutated white skin genes. As the African races get to travel and intermarry more, pale skin should also become darker.

By the time the true blonde (and redhead) becomes extinct, blonde-gene therapy should be around for those who need it. Successful gene therapy is now possible in animals to change their hair colour. For the next few years, however, the power of the chemical blonde will continue to dominate the Western world – and be a historical tribute to our Viking ancestors.

Ageing and longevity

S was 65 years old when, for the first time since her last pregnancy, she had to visit her doctor. She was troubled by a number of minor but irritating complaints, including haemorrhoids, back pain and episodic hair loss. After a full check-up, she found that her cholesterol levels were high, her blood pressure was raised and X-rays of her spine showed arthritic changes. Everything seemed to be going wrong at once, after she had been so healthy for so long. Her handbag now resembled a chemist shop. She also noticed her skin getting drier and more wrinkled, despite spending a fortune on creams and cosmetics. Her mother had similar problems at around the same time – and S wondered why this should be.

The genes in our bodies have been faithfully copied and passed on for the last 5 billion years, yet our bodies are supposed to last only a biblical three score years and ten. Why don't we live for ever? Our bodies are pre-programmed to self-destruct. But why do most of us fall apart slowly rather than suddenly drop dead? Our successful ancestors, who survived all other perils, had all their children by the age of thirty and spent maybe twenty years looking after them – perhaps adding a further ten years as grandparents. By the ripe old age of sixty their genes had no further need of their temporary owners and they could allow them to wither and die.

Ageing is related to the way our cells divide. After fifty doublings, our bodies are created from a fertilised egg, but certain tissues, such as skin and the lining of our arteries, keep dividing throughout life to repair themselves. After several copies, the 'photocopies' would normally start producing a few errors at the edges – rather like missing the first line of a page.

Genes called telomerases prevent these errors. Their role is to add some garbage DNA before every copy is made – like extra space around the page – so nothing important gets missed off. Unfortunately, telomerase genes get switched off after a certain number of copies are made and our cells auto-destruct. We slowly lose slices of our DNA at an average rate of around thirty base pairs per year. Some people inherit particular genes, which produce longer and better telomeres – potentially allowing a longer lifespan (although not yet proven). Twin studies have shown the length of the telomeres between different individuals varies and is around 80 per cent heritable.

Other genes involved in ageing include those that defend DNA and tissues against damage (antioxidants) and repair DNA when things go wrong. Smoking is a good example of a factor that increases damage to many tissues and puts pressure on cells to divide and repair, using up their supply of telomeres. This causes premature ageing in different cells of the body (e.g. wrinkles, hardened arteries). When the DNA repair system works too well, and stops the cells auto-destructing after their normal lifespan, the risk of cancer increases. Thus, at a simple level, premature ageing may be preventive for some cancer processes, because, if the cells die early, they are less likely to become cancerous.

Another gene that may be involved in ageing is the Ddc gene, which synthesises depamine and serotonin. It has been found in fruit flies and could add an equivalent of five human years to life.

The oldest documented human was a French woman, Jeanne Calment, who lived to 122 years and died in 1997. Her lifestyle and diet were studied in great detail to see if they could reveal the secret of life. She had one child and her 'healthy' lifestyle consisted of a French diet of butter, fatty foods and garlic, as well as smoking cigarettes until the age of 117. (Why did she bother giving up?) George Burns, the famous comedian who died at a hundred, was famous for his poor lifestyle and cigar smoking. Some people believe that the rare group of long-term smokers who remain healthy have special genes producing high levels of protective chemicals (heat-shock proteins).

Reaching a hundred, which now occurs in 3 per cent of women, seems to be more related to your genes than your lifestyle. If you have a sibling who has lived to be a hundred, your chances of doing the same are increased fourfold. Jeanne Calment had 32 great-great-grandparents, mostly shop-keepers from Arles in the eighteenth century. Apart from a few who died of accidents, all were long-lived. Her brother died tragically young – just before he could reach the age of 98. Anecdotally, twins confirm the importance of genes on longevity – Gin and Kin, a pair of Japanese identical twin sisters, were the oldest in the world at 107. Other twin studies in Scandinavians have recently confirmed a genetic effect on longevity, with a heritability that approaches 50 per cent in the over-seventy-year-olds.

One gene, called Apo E4, is involved in the way fats in the body are absorbed, and is related to a number of common age-related diseases, including heart disease, osteoporosis and dementia, leading some people to think this may be one of a large family of genes involved in the ageing process. In a study of 75 Japanese centenarians exploring diet and physical activity, one of the most important factors was the Apo E gene, in particular the E2 variant, which controls lipid and HDL levels.

Why genes controlling aspects of nutrition should influence ageing is puzzling – but is supported by experiments on laboratory animals. These show that their lifespan is significantly increased if they go on restricted diets, which limit their height and weight. The drawback is that while on a diet the animals can't reproduce. In effect, the body is put into holding mode, and metabolism is slowed down and ageing is stopped, until nutrition is restarted and the body has enough resources to make reproducing worthwhile. The role of height may also be crucial. It is well known that dog longevity is inversely proportional to size, and this is true for most animals studied. Most human centenarians are small in stature and the longest-lived population, the Okinawans, are one of the shortest, with their centenarians being 4 inches (10 cm) shorter than 75-year-olds. Studies of many populations have shown a clear relationship between short stature and longevity. For example, of 3,200 deceased US baseball stars, the average age of the short ones (5 feet 3 inches, or 162 cm) at death was 81 compared with the tall (6 feet 4 inches, or 193 cm) of only 69 years.

These observations provide clues as to why some humans age fast and die young, and others live to be a hundred. Studies show us that major changes in ageing can be achieved with just a few genes. The worm *C elegans* can now be manipulated genetically with the gene SIR2 and many other genes to increase the lifespan up to fivefold. The gene senses scarcity in the environment and can slow down the ageing process and increase resistance to infection or environmental damage. The downside is a corresponding reduction in ability to reproduce as frequently.

The genes we inherit appear to provide us with two evolutionary survival strategies. The first is to put most of our body's resources and energies into the early parts of our lives, ensuring survival and reproduction with as many children as possible. There is not much energy left to protect the body against later infections, deterioration and gene damage. The other strategy is to conserve energy, be less aggressive in pursuit of mates and having children and have plenty of reserves for fighting disease and decay later. There is now evidence to support this. An extensive study of 33,000 individuals from genealogy records of mainly the British aristocracy (15 per cent were also French and German) over several centuries showed clearly that the married women who lived the longest had the fewest children. Women living to over ninety had a 50 per cent chance of being childless compared with only 30 per cent of women dying before they were fifty. Those living longest were also likely to have their children later in life (although, as we've seen, they were more at risk of breast cancer).

There is thus a clear inverse relationship between reproduction and lifespan. In most countries, lifespan is increasing, due partly to better nutrition, but also perhaps to greater chances of our offspring's survival, which allows us to invest less in producing them. New studies suggest that, with improving nutrition and control of infectious diseases, there is no clear natural limit to our lifespan. So, if you want to live to be a hundred, don't waste too much energy trying to grow, or finding a mate: instead, have a minimum of children and have them late in life (if you want your genes to survive), and retire to Miami or Marbella with a good pension scheme and plenty of time to improve your golf.

Biological clocks and genetic grannies

L was aged 32 when her older sister (aged 36) told her she had been diagnosed by her doctor as having an early menopause and was suffering from sweats and flushes. This worried L, as, unlike her sister, she hadn't yet had children herself. She suspected that her sister's unhealthy lifestyle (she was a heavy smoker) was to blame. Although she was in a steady relationship, her job was going well and she wouldn't have chosen this moment to have children. Her mother had needed an operation to remove her womb (hysterectomy) for fibroids when she was 42, and couldn't tell when her natural periods would have stopped. L thought no more about it, but was shocked when, as with her sister, her periods began to stop five years later.

The average age at menopause is around fifty in developed countries but varies widely. It marks the end of female fertility, when the ovaries cease functioning. Female fertility gradually reduces before this time, from a peak in the early twenties, with a marked reduction after the age of forty. The end of reproduction is accompanied by a drop in the levels of oestrogen that in over half of women leads to flushes, night sweats and mood changes. These symptoms can last a few months up to two years. After the menopause, the rates of certain diseases, such as osteoporosis, increase and heart disease rates become more similar to those of males.

Twin studies from our group have shown the timing of the menopause to be 60 per cent heritable. The age of menopause of your close relatives is a good indicator of your own timing – with the menopause of most women being within five years of their sister or mother. The symptoms of the menopause also appeared to be more similar in identical twins than nonidentical twins. Cigarettes are the only proven lifestyle factor, but bring forward the menopause by only one or two years.

About one in four women now undergoes operations to remove the womb (hysterectomy), usually around the age of forty. Strangely, this is commoner in Californians and the wives of doctors, who have a 50 per cent chance of surgery. This operation also shows a strong genetic basis related to the growth of fibroids in the womb or excess bleeding. If L hadn't had a menopause early in life, she would have had a strong (50 per cent) chance of needing a hysterectomy in her forties, like her mother.

Humans are nearly unique among mammals for having a menopause, the others that do being the pilot whale and the elephant. The females of most other species have a slowly diminishing fertility until they die. One evolutionary explanation for having a clear time point when the ovaries fail is the 'grandmother menopause theory'. As a species, humans have evolved for quality, not quantity, in child rearing. Producing endless numbers of children every three years would have led to most of them dying, particularly without a guarantee of support. Once a woman had produced a few children, her best genetic option to ensure that her genes survived was not to have more, but to help her grandchildren. In the Pleistocene, the ability for the group to keep moving to find food must have been crucial and would have been hampered by a cave full of screaming toddlers. An extra pair of experienced hands to carry infants would have greatly aided the survival of the next generation – and the continuation of the menopause genes.

Brittle bones

C was 55 when her mother first broke her hip. She remem-bered later how she'd also broken her wrist in her fifties and later developed a stooped back, or 'dowager's hump'. A friend told her that this might be osteoporosis. A screening test (with a bone-density scan) confirmed she had thin bones and was at threefold increased risk of fracture. She took preventive treat-ment and successfully avoided her mother's fracture problems.

Osteoporosis (or brittle-bone disease) is a good example of a very common heritable illness, affecting about one in four women and one in twelve men in their lifetime. The strength of your bones is determined mainly (75 per cent) by your genes. The risk can be passed down from your mother or father. If your mother had an osteoporotic fracture – hip, wrist or spine – your risk is increased threefold, making it quite likely you will have problems later. Diseases in this common and heritable category include heart disease, diabetes, strokes, high blood pressure, high cholesterol, arthritis, migraine, obesity, haemorrhoids, hernias and stomach ulcers. As they are so common, your chances of also being affected are quite high.

Fracture rates are still increasing in most countries, showing the effects of our ancient genes interacting with our modern environments – sedentary lifestyles and low calcium intakes. Treatment (with bisphosphonates or oestrogen-like drugs), if started in time, is now very effective and can reduce the chances of new fractures by 50 per cent within a few months of starting therapy.

That osteoporosis is related to genetic mutations is shown by the fact that it is virtually unknown in people from pure

African origin, and is most common in north Europeans and the Nordic races. One explanation for this is that as man moved out of Africa northwards into climes where there was less natural sunlight, a problem developed as children's bones failed to grow properly (a condition called rickets) without the sun, a major source of vitamin D, essential for bones. The body adapted and, as we discussed earlier, evolved by having lighter skin, which, with the help of other gene mutations, absorbed vitamin D more easily and prevented childhood problems. Unfortunately, the side effect of this successful past adaptation works less well in an environment of little exercise, indoor living and poor diet, resulting in osteoporosis in older life. Successful public health campaigns in many countries have started to reduce our exposure to sunshine. While this has been undoubtedly helpful in reducing the incidence of skin cancer, the side effects are vitamin deficiency, increasing risks of osteoporosis, rickets, and a number of cancers such as colon and prostate cancer, which, overall, might be a greater public health problem.

Recently, a few genes have been found that influence bone strength, including the vitamin D receptor gene, which affects the way vitamin D and calcium are absorbed, oestrogen receptor genes and genes controlling fat and cholesterol (LRP5 genes). Milk is a major source of calcium in Western diets and has been shown to protect against thin bones later in life – surely a classic environmental factor. But we have discussed previously how some people lack the gene mutation (LCT) to help them drink milk as adults. One in four women has this mutation, which has been recently related to both aversion to and reduced intake of milk, as well as leading to an increased risk of fracture. A few families have been found with 'super-strength' bones that had these special cholesterol (LRP5) genes. No one in these families ever suffered a fracture – even when hit by a bus. As with all genes, where there's an upside there's usually a downside and these rare super-strength families may get cancer more often than average.

Entrepreneurial genetic companies (such as www.
genovations.com) are now offering a genetic screening
service for osteoporosis risk (using mouth swabs) to test for a
panel of six genes for around £200 ($300). Although the
situation should improve, no one knows currently how to
interpret the genes in terms of advising on traditional
treatments and the results are likely to be no better than
analysis of hair samples or looking at tea leaves.

Family back problems

*M at age 37 was a plumber by trade and never very fit. But,
being good at DIY, in between jobs he decided to build an
extension and loft conversion. He hurt his back while lifting
and developed sciatica (pain down his legs). He rested for two
weeks but the problem recurred, and he was never again free
of back pain and never finished his extension. He eventually
had to take early retirement from his job and retrained for an
administrative position. He had, prior to his injury, suffered
short episodes of mild back pain in the past lasting a few days
at a time. His father and uncle had both suffered from disc
problems and sciatica early in life – forcing them to change
careers – but he'd never considered it was heritable.*

Back pain and disc problems cost the UK alone an estimated
£12 billion. It affects the majority of us at some point in our
lives. At least one in three people suffers long-term problems
and one in six suffers severely, causing disability. It is the
leading cause of lost work time in developed countries and
one of the major claims and financial drains on employers
and insurance companies. For years the dogma (without
much evidence) had been that the causes were purely
environmental – due solely to occupation and lifting weights.
Despite many studies over the last fifty years comparing
sufferers and normal subjects, no clear risk factors emerged.
Recently, back-pain susceptibility was studied using twin

studies and, against expectation, found to be genetic and 60 per cent heritable. The results relate to a threefold increased risk if you have a first-degree relative with the condition. Given this is so common anyway, the chances are very high indeed.

Determining how genes could be responsible for such a common problem is hard to grasp. Genes could affect the anatomy of the spine, affecting the subtle variations in the way the muscles, bones and cartilage work together, altering posture. Another way might be at the biochemical level: genes might control the laxity of the collagen fibres that make up the spinal ligaments or the amount that discs can act as shock absorbers without bulging out and pinching nerves. The third way is via the brain (the thalamus and the amygdala), influencing our pain thresholds and other factors – notably anxiety and depression.

Pain responses and thresholds are highly variable in humans and clearly heritable in laboratory animals. Human twin studies have shown a range of reported painful symptoms to be influenced by genes – headaches, stomach, elbow, hand and shoulder pains. A few rare families were discovered who have genetic mutations that make them exquisitely sensitive to the slightest change in pressure – such that touch can induce pain. Other rare families have a mutation in a gene (NGF) that makes them insensitive to pain. Less is known about pain responses in normal people and a number of genes are likely to be involved. Recently, one gene (COMT) has been found to control how the chemicals dopamine and adrenaline cause a sensation of pain in the brain. Individuals with the COMT mutation experienced significantly more pain and anxiety when given a painful slow infusion of a liquid into the cheek than those with the common version of the gene.

Neck pain was also found in our twin studies to be common, affecting about one in four people and clearly genetic with a heritability of 60 per cent. However, on examination of the changes these people had to the structure of the neck using MRI scans, no relationship between pain and anatomy was found. Indeed, it seemed that the most important factor for neck pain were genes controlling anxiety. In back pain and in other painful conditions, genes for anxiety and mood clearly play an important role in how people respond to or put up with pain. While most of us experience some neck or back pain in our lives, only a minority of individuals develop long-lasting (chronic) problems due to sensations of continuous pain and resulting disability. It was thought that much of this was explained by environmental or social factors. However, one of the most consistent findings in these sufferers is that they express high levels of fear and anxiety in response to the initial problem. We discussed earlier how fear reactions are variable and heritable and that genes influencing serotonin (such as COMT) regulate 'pain thermostats' in our brain, so that all signals of pain become exaggerated. People, who can face their initial problems head on without this fear reaction keep their pain thresholds stable and rarely develop these disabling symptoms. Although individuals are genetically predisposed, targeted drug or behavioural therapy could help these sufferers in the future.

What should M, with his strong family history of back problems, have done? It's probably impossible to avoid back problems altogether, but he can reduce the risks of permanent disability by keeping himself fitter, thinner, more flexible and with knowledge of how to lift without injuring himself. People with strong family histories of back problems should probably think twice about high-risk occupations such as construction or nursing, which have high rates of early retirement because of back problems. If they do suffer an injury, a positive 'fearless' approach is crucial to recovery.

Self-destruction

K's mother was diagnosed as having rheumatoid arthritis, a disease that causes painful, swollen, stiff joints, particularly in the hands and feet. The specialist had explained that people with this autoimmune disease (where the body attacks itself) have a one-in-three chance of becoming disabled – but new drugs can help considerably. K's mum gave her daughter a leaflet that described all the symptoms. It explained that it often ran in families, was common in females and carried a fourfold increased risk in relatives. Naturally, K was very worried. She went to see her doctor, who didn't know the details of the inheritance of the disease, but agreed to run some blood tests.

Autoimmune diseases affect around one in a hundred to one in ten thousand of the population and are less common than the examples we have been discussing. In general, they all have a genetic component, on average with heritabilities of around 60 per cent. Examples include multiple sclerosis (MS), rheumatoid arthritis, colitis, lupus, myasthenia gravis and epilepsy. Although the risk of your getting the disease is proportionally increased if you have an affected relative, you are much less likely to be affected because autoimmune diseases are rarer, and, like the commoner diseases, are influenced by a large number of genes. For example, if rheumatoid arthritis occurs in your mother and affects one in two hundred people normally, even if your risks are increased three- to fivefold, you still have only a very small chance of being affected (one in fifty). Some so-called autoimmune diseases are interrelated, so a family history for one can influence your chances of getting the other. Examples are thyroid disease, pernicious anaemia, insulin-requiring diabetes and rheumatoid arthritis. However, the absolute risks for most people will still be low.

Some autoimmune diseases have genes that act as markers for the disease and are present in most people with the condition. The best example is a gene (B27) on chromosome 6 from the HLA system of cell recognition. This genetic marker is carried by more than 90 per cent of patients with a disease of the spine and joints called ankylosing spondylitis, compared with only 8 per cent of unaffected people. Unfortunately, it is not much good as a screening test as fewer than 10 per cent of people with the gene develop the disease. Similarly, a gene (HLA DR4) occurs five times more often in people with rheumatoid arthritis than expected, but is still not good enough to use as a clinical test.

Some of the autoimmune diseases, such as rheumatoid arthritis, are fairly recent in our history. Rheumatoid arthritis was not described before the seventeenth century and its occurrence seemed to increase in the 1950s and 1960s, before it begins to reduce again recently, suggesting some change in environmental triggers, such as viruses, pollution or diet. Most of the autoimmune diseases are likely to be caused by an underlying genetic tendency plus some other triggers at a crucial time of life. We've seen before how genes don't set out to cause illness, and diseases are usually a side effect of another process or benefit. The fact that autoimmune diseases are much commoner in women may be a side effect of their having better protection than men against infections and parasites such as intestinal worms – hookworm, for instance. It may also be due to a temporary lack of control by the immune system, which has to switch itself off during pregnancy to avoid destroying the 'foreign' foetus.

The genes themselves may protect against other diseases. One example is that rheumatoid patients strangely never get schizophrenia (a cause of psychosis and mental breakdown) and cope with the mental stress of their disease well. An unusual benefit of an autoimmune disease came to light in a male patient who had ankylosing spondylitis, which caused him to have a stiff, rigid spine. He became the US middle-weight arm wrestling champion. When his doctor told him of the reason for his back pain, he refused treatment, saying that having a rigid spinal support as leverage gave him the crucial edge over opponents.

Relatives behaving oddly

Just after midnight, H was woken by the police, who informed her that her older sister, T, had committed suicide with a massive overdose of tablets. T had been suffering from depression for years with bad episodes following the birth of her two children. T hadn't believed in medication and had resisted her GP's and psychiatrists' attempts to allow them to treat her properly. Several months later, H's husband became worried about his wife's state of mind and made an appointment to see their doctor together. He wanted to know H's risk of suicide. Before the planned consultation, H herself took an overdose of sleeping tablets. Her husband found her in time – she survived and later slowly improved with long-term treatment

All the common mental and psychiatric disorders – such as depression, anxiety, neurotic and obsessional behaviour and manic and postnatal depression – have a clear hereditary basis. Twin and family studies have confirmed that on average well over half the influence is genetic – with other influences being random life events or exposures. In general, studies have found little or no effect of family environment. These conditions often lead to problems with functioning in normal modern life (known by the misleading phrase 'mental breakdown'). All these disorders are related to abnormal levels of brain chemicals (dopamine, serotonin and others) that are regulated by genes. As yet we know little of the principal genes responsible, but we are getting better at improving the symptoms with drugs.

Depression is one of the commonest disorders in modern society and, depending on how you define it, affects up to one in ten women and slightly fewer men. It is one of the commonest causes of lost working time and is the most important disorder in insurance payouts. There is a wide spectrum of disease with some people being prone to cycles of low mood and others experiencing short bursts of elation and energy (mania) in between bouts of depression. The spectrum of so-called 'affective disorders' is believed to have common causes. The risks of it are increased three- to ninefold fold in first-degree relatives – leading to a greater than one-in-three chance that they themselves will be affected. The risks may be higher if the disease is more severe or occurs at an early age in the family (before twenty). An identical twin with an affected twin will develop depression only 50 per cent of the time on average – so environment is clearly also important.

For years we've known that adverse life events are related to depression, and curiously a tendency to experience these events or find yourself in stressful or bad situations is also partly genetic. Why some people cope with stressful life events and others get depressed was unclear until recently. A New Zealand study showed that variations in the promoter region of the serotonin transporter (5HTT) gene were responsible. Subjects faced with four or more life events had a one-in-three chance of major depression before the age of 26 if they had the 'at risk' genotype, compared to a less than a one-in-six chance if they didn't. Drugs correcting the brain's imbalance of dopamine and serotonin can reduce many of the symptoms and resulting problems. Prozac, one of the world's best-selling drugs, blocks the reuptake of serotonin, allowing it to remain longer in the brain, and make us happier. There is currently only weak evidence implicating certain genes – such as the dopamine (DRD3 and DRD4) genes – so genetic testing is still a long way off.

Suicidal tendencies have also been shown to be genetic from family, twin and adoption studies. Strangely, this genetic connection may be independent of any other major disorder such as depression – so it can be inherited on its own. Several studies of suicide or attempted suicide have shown an association with low serotonin levels and the serotonin transporter gene, which is unique to apes and man. The large differences in suicide rates between countries – such as the high rates in Nordic countries and

Japan – is likely to be explained partly by cultural factors and partly by genetic differences.

Other mental disorders such as schizophrenia are also strongly genetic, but are much less common, with a lifetime risk of less than 1 per cent. The risks to first-degree relatives are raised somewhat because of the high heritability (80 per cent), and around one person in ten is affected. It usually occurs early in life and can have a devastating effect on families as sufferers lose touch with reality, become paranoid and hear voices. Identical twins can often differ in disease expression or symptoms. The infamous London gangsters the Kray twins were both violent, but only one (Ronnie) had obvious schizophrenia. Again, adoption and twin studies have not shown any effect of family environment – so we can't blame their mother or their East End cockney upbringing. More schizophrenics are born in winter than summer, so it is possible that, if the mother has a viral infection during pregnancy, this might trigger the genetic tendency. A few genes have been implicated in schizophrenia, but have only weak effects, suggesting there may be many different ways to acquire it – potential genes so far include the dopamine receptor (DRD3) and serotonin receptor (5HT2A)

It is believed by some, but difficult to prove, that the genes for these disorders must have conferred some benefits on our ancestors. Patients with schizophrenia, and to some extent their relatives, appear to have fewer viral infections and have increased resistance to pain or to developing autoimmune diseases (for example, rheumatoid arthritis). Having psychotic episodes or visions may have improved the status of some of our ancestors – some psychiatrists believe Joan of Arc was a good example. Many past geniuses, such as Issac Newton, may have been mildly schizophrenic, and John Nash (of the film *A Beautiful Mind*) suffered but still managed to win a Nobel prize. A little dose of these genes may have been very useful for creative thinking. The disease ran in the families of the Einsteins, the Jungs and the Joyces.

The benefits of low mood or depression in our ancestors may have emerged from different directions. Communicating a need for help, giving us a way out in conflicts with dominant figures, offering an easy way out of commitments to unreachable goals, or avoiding situations likely to result in danger, injury or wasted effort would all be useful in certain situations. Postnatal depression soon after birth is one common situation where being depressed may be a defence mechanism to seek help from others (and shown by twin studies to be also genetic). Some believe this behaviour may have developed to counter an impulse to kill the newborn baby (infanticide), which we believe was reasonably common in our ancestors. A related disorder to depression is bipolar disorder (manic depression), which is also highly heritable. Highly fluctuating responses to brain hormones cause swings in mood from euphoria to depression. The advantages from being on the 'up' side are seen from studying famous sufferers who managed to achieve great things. Their disorder temporarily gave them extra energy, insight and creativity. Those believed to have been affected included Winston Churchill, Lord Byron, Van Gogh and Sylvia Plath. One gene uncovered so far is the GRK3 gene, which makes sufferers overrespond to the effects of dopamine – causing excitement and stimulation for a few weeks or months, followed by weeks or months of depression.

Losing your mind

P was 45 and had always been emotionally close to his mum, who was now 75. Until recently his dad had looked after her, but he'd just died from a stroke. She couldn't cope on her own and P had made a flat for her in the basement of their house. For the last four years her memory had been deteriorating and she'd been diagnosed as having early signs of Alzheimer's. This was very frustrating for her, as she'd been a university lecturer. She could no longer read more than a page at a time and writing more than a sentence was an effort. Since she started taking the drug tacrine she had some better days, suggestive of recovery, but soon relapsed a few days afterwards. One day P came back from work and his mother wasn't at home. He searched the local shops and streets to no avail. That night the police brought her back – she'd been wandering lost in a park five miles away, and couldn't remember her name or address. P began thinking of his own risks. He knew his aunt had had similar problems. He made up his mind to see his doctor to discuss getting some tests on himself – and asking how he could prevent it.

Dementia means literally losing your mind. It specifically refers to a disease process of losing your memory to an extent that affects your life – so is much more than just occasionally losing your keys. There are at least 55 diseases that can result in dementia, the commonest being Alzheimer's and vascular dementia. Alzheimer's is the commonest cause of dementia in the elderly: it affects 5 per cent of people over 65 and 40 per cent of those who make it past 90. It starts as a subtle memory loss then gradually worsens until most brain functions deteriorate. A close inspection of the brain has shown that the brain cells die and are surrounded by a network of plaque and tangles, made up of excess amounts of key brain proteins (ß amyloid and tau).

Dementia patients are understandably difficult to study for research, as they are old, forget their appointments, and have problems with questionnaires and consent forms. Nonetheless, several determined investigators using twin and family studies have shown that Alzheimer's and the other dementias are strongly genetic. Identical twins get the disease if the other is affected around 80 per cent of the time. A risk to first-degree relatives of developing Alzheimer's by the age of 85 is 25–40 per cent, and relatives will get it earlier in life. The only environmental factors shown to increase your risk are age, being female (which you can't do much about) and limited years of education. This suggests that the more information you fill your brain with, and increase your neural connections, the longer it takes for the effects of Alzheimer's to be seen. Head injuries are the other major risk factor, which can double your risk. Recently the high risk of soccer players who repeatedly head the ball has been highlighted by the premature death of the English footballer Jeff Astle and others, prompting a call for this to be called an occupational injury and soccer to be considered a high-risk sport, like boxing.

Gene finding in the dementias has been very successful. In many of the very rare and severe forms that run in families, such as Huntington's chorea, the key gene has been identified. This is also true in the rare early-onset Alzheimer's that occurs aged forty to sixty, where the gene (APP) making the protein amyloid is abnormal and produces excessive amounts, which gets laid down in the brain. This gene is on chromosome 21 and explains why people with Down's syndrome (who have an abnormal chromosome 21) get similar brain problems in later life. Two other genes (PS-1 and PS-2) have also been found to increase amyloid plaques. While the genes (over seventy-two variants from three genes) explain more than 50 per cent of early-onset disease, this accounts for only 2 per cent of the common form of Alzheimer's, where there is more to the story.

The strange and ubiquitous gene on chromosome 19 that clears fats from the body – ApoE – is again involved. It reduces the clearance of the proteins amyloid and tau from the brain, therefore allowing more plaques to form. If you have a single ApoE4 variant, your risk of Alzheimer's is increased threefold, and with the double dose your risk increases to thirteenfold. Unfortunately, there are many exceptions: one in three people with Alzheimer's don't have the variant ApoE4 gene and up to 50 per cent of people with the double dose don't get the disease by age eighty. Other genes are clearly involved.

As yet there are no cures for these dementias, although some of the early symptoms can be slowed down somewhat in about 50 per cent of patients using anti-cholinesterase drugs (tacrine, galantamine), which increase communication between the healthy cells in the brain. A major dilemma is whether to screen relatives for the condition. In the early-onset diseases (like Huntington's) predictive genetic tests are much more accurate than for late-stage diseases, but in both scenarios there is no cure. In these situations most people prefer not to know their fate. More difficult is the decision to have screening tests before having children. Some people prefer to not have children rather than face the uncertainty of knowing.

Finding a possible biological or evolutionary explanation for dementia is highly speculative: It is one of those problems of age that most of our ancestors didn't have to worry about. The amyloid-like proteins (pre-senilins) probably have a key evolutionary role, as they are found in simple creatures like worms and can affect egg laying. One clue might be that the amyloid protein has a protective role and may help fight infections – so having excess might have been a good idea. If you are planning on living a long time, it's likely you will develop some degree of dementia. The best defence seems to be to die young or keep your brain busy.

Infectious immunity

Two students, T and J, took a gap year after finishing school and went on a three-month trip together to India. They ate the same food, suffered the usual stomach problems, lost some weight – but had a great time. On returning home, J found it difficult to regain the weight he'd lost, while his friend T (on the same fast-food diet) rapidly became chubby again. J developed a cough, lost more weight and was tired all the time. He was very ill by the time tuberculosis (TB) was diagnosed from his chest X-ray and his sputum sample. Like his friend T, he'd been vaccinated several years before while at school, and neither he nor his anxious parents could

understand why he and not T had developed TB. J improved
slowly on treatment over the next six months.

At least a third of the world population is estimated to
be infected with a stubborn little bacterium called
Mycobacterium tuberculosis, which causes TB. About
8 million new cases occur worldwide and 2–3 million people
die every year of it. Many famous people have died in the past
from 'consumption' (such as Keats and Robert Louis
Stevenson). TB in the West has become more common
recently due to increases in immigrants from countries where
it is endemic; increased world travel; more homeless people
in crowded conditions; increasing numbers of the elderly; and
more AIDS patients. Although the trends have been reversed
in the last few years, more than eighteen thousand cases of TB
were reported in the United States in 1998. There is only one
vaccine (called BCG) and it has variable effects in different
populations, ranging from zero effectiveness to 80 per cent,
dependent on the individual's genes and environment.

Twin studies have confirmed that the body's ability to
defend itself against TB via the immune system is highly
heritable. Large differences are seen in infection rates and
clinical reaction to the bacterium (and vaccines) between
individuals and races. Genetic effects are believed to work at
a number of different levels. First, they control the level of
bacteria that stay in the body; second, they determine
whether these bacteria cause clinical problems; and finally
they affect the severity of the condition (whether you die).
Different genes control each of these processes.

Some rare mutations in immune genes can cause children
to be infected by usually harmless forms of TB bacteria – such
as those found in fish tanks or in vaccinations. These genes
are too rare to explain the different responses between people
to common infections. In adults, a key gene (NRAMP1) seems
to be one that alters the function of the white cells that are
designed to engulf and destroy the bacteria. In one study of an
epidemic in Canadian Indians, having only one copy of this

gene increased the risk of getting clinical TB tenfold. Other genes undoubtedly play a role. Those implicated include the HLA family, the TRL-4 gene, which limits TB growth in the lungs of animals, and the vitamin D receptor gene, strangely linked to a number of other diseases.

In our Western ancestors, having variants of genes that protected against TB (such as NRAMP1) would certainly have been an advantage. The first known cases of TB in the UK are from around 2,300 years ago and those in Italy and Egypt from around 6000 BC. In later times, epidemics in Europe accounted for around 20 per cent of deaths, particularly in crowded towns and cities. Another such gene variant that offers protection against TB is the HEXA gene, common in Ashkenazi Jews, who, living in confined ghettos, were particularly at risk of TB. One in thirty Jews carries a single copy of the gene that is harmless – until two carriers mate and produce a child with a rare but severe neurological disease called Tay-Sachs. The genome of the mycobacterium has now been unravelled and should allow us better vaccines and more targeted treatments against the defences of this stubborn enemy.

The big killer diseases such as smallpox, tuberculosis, measles, whooping cough and influenza are relatively recent in our history, having affected man only since he started domesticating animals around ten thousand years ago. Many people died as a result, particularly in crowded emerging cities – but the survivors were genetically tougher and had developed partial immunity. The incredible speed of the conquest of the Americas in the sixteenth and seventeenth centuries over the vast numbers of indigenous peoples was mainly due to these diseases that the European invaders imported – to which the natives had no natural genetic defence.

Dropping dead

N was in his forties and had children of his own. Many years ago he had left his native Wales to work in the Caribbean as a policeman. One day he was looking at family photo albums, thinking of the time he would return home. He was prompted to remember his roots and he started to write out the sad story of the family. His mother had died suddenly aged thirty while he was too young to remember the details. His brother K had died aged sixteen in strange circumstances while swimming with him in a river and was thought to have had a fit. His sister Y had also died suddenly aged nineteen, after a series of blackouts. His dad and he had escaped. When he looked even further back, he found more early deaths on his mother's side. The illness in the family had been given the nickname the Gorry family curse – as no one had found any clear medical reason for the deaths. N wanted to know more about the risks to his children and to see if anything could be done to prevent it.

When N started doing more detective work he found cousins and relatives scattered all over the world, from Canada to Australia, most with the same story – of tragic early unexplained deaths. Eventually he came across some cousins in Australia who held the answer. One girl was having unexplained blackouts and her doctor finally performed an ECG trace of her heart. The results clearly showed a disease called long QT syndrome, an abnormality of heart conduction in times of stress that can cause the heart to stop. The disease is often hard to detect but can usually be prevented by simple medication (beta-blockers) and is inherited. N constructed extensive family trees and contacted as many relatives as he could find and obtained their DNA. Unfortunately, about twenty out of sixty had died prematurely, some just weeks before. From this, a lab found key markers for the mutated gene and this was used to test family members: 50 per cent of them were free of it, including, to his relief, N and his children. However, several members of the family were diagnosed in time and owe their life to his detective work.

Long QT is a rare condition affecting one in ten thousand people with many hundreds of different gene mutations causing roughly the same disease pattern, making it difficult to diagnose genetically. The mutation affecting a single protein that was crucial for the electrical response of the heart arose by chance in a single relative of the family several generations before, and has been passed on to 50 per cent of his or her children ever since. Usually genetic mistakes like this don't last very long, as the individuals with the gene die before they can reproduce, but in large families like the Gorrys, enough children can be produced to continue it for several tragic generations. This real-life story illustrates clearly how investigating your own family history could save your life.

Will I die young of heart disease?

M had just turned fifty and on altering his life insurance he had to answer a number of difficult questions about his family history. His father had died young, at 57, of a heart attack, but his mother, aunt and uncles on both sides were still going strong in their eighties. He had a brother five years older who seemed fine. A few weeks later he was told his premium would be higher, as he was considered at risk of dying early (which is what insurance companies worry about most). Through the insistence of his wife and children, he was eventually persuaded to have a medical check-up – fully expecting to have a clean bill of health. When younger he had smoked too much and on the whole had a poor diet with a penchant for meat and fatty foods. Recently, he had started cutting back on these excesses and felt healthy. To his surprise, his doctor told him he had high blood cholesterol levels and needed to take action if he was to avoid the same fate as his father.

Heart disease is the major killer in Western countries and accounts for about a third of all deaths. Heart disease is a term for the clogging up of the arteries, leading to sudden blockages (heart attacks), spasm of the arteries (angina) or long-term lack of blood supply (heart failure). Unlike long QT syndrome, it is due to multiple environmental and genetic factors.

Large twin studies of over more than twenty thousand pairs have shown death from heart disease to be highly heritable with an overall estimate of 57 per cent, which increases with early deaths. The risk of death due to heart disease in an identical twin if the co-twin died before 65 years was around tenfold and around threefold for fraternal twins. The risks for first-degree relatives are likely to be around two- to threefold. A pair of Australian twins who lived all their lives together illustrated this well when they celebrated their seventieth birthdays together at a Perth restaurant and dropped dead of heart attacks within thirty minutes of each other. Another identical twin from England suffered a massive heart attack and was saved by cardiac surgery to find that his twin brother had died from a heart attack on the same day hundreds of miles away.

Rates of heart disease have increased fourfold over the last eighty years in developed countries, but it remains a relatively rare cause of death in Africa and Asia – in China fewer than 1 per cent of deaths are due to heart disease. It is caused by multiple intermediate factors, including high cholesterol and blood fats, high blood pressure, sticky blood and lack of elasticity in, and inflamed lining of, blood vessels. Genes influence all of these intermediate factors. At the population level it is strongly related to Western habits such as obesity, smoking, fat intakes and sedentary lifestyle, and related to social class and education level. It is also commoner in males, particularly at young ages. In the 1970s the highest rates in the world used to be in Finland, but a major public health campaign has started to change the diet of the population and they have cut back on high fatty foods and dairy produce and are eating more fresh fruit and vegetables. They are no longer top of the death league (Scotland has that dubious honour). These changes can't be due to genes alone and show the importance of recent changes in our diet.

However, people cope with fat in the diet differently, and genes are responsible for this variability. Some of our ancestors must have eaten large quantities of meat. We believe they may not have had meat every day (as fridges and freezers were not available), but, when it was on the menu, it's likely they ate as much as they were physically able. So why didn't they get heart disease? The large variety of fruits and nuts they ate may have been one reason – but also the nature of the meat itself. The domestic cattle, beef or lamb we eat today is around 20–25 per cent fat, whereas the free-range game our ancestors caught and ate (such as antelope or venison) is only around 5 per cent fat – a difference likely to be crucial.

We don't know exactly what caused M's dad to have a heart attack, but it's quite likely that genes for high cholesterol played a role. M's brother S reluctantly had a blood test and was also found to have a high level of blood cholesterol. In this way blood cholesterol is acting indirectly as a genetic test, as it has a heritability of around 70 per cent. M and his brother were both put on a low-fat diet for three months, but their blood levels only improved slightly and they were started on cholesterol-lowering drugs, called statins. These reduce mortality from heart disease by 50 per cent, but can cause depression and very occasionally suicide – as cholesterol is an essential component in many hormones and brain chemicals. Their risk of a cardiac problem while on medication will still be around 50 per cent higher than average, but considerably better than the threefold risk they started with.

High heritabilities and the genetic influences on certain diseases have important consequences for the insurance industry. Insurance companies have a wealth of unfortunate but basic statistics to help them assess risk and come out slightly ahead. They are increasingly relying on family-history questions (such as 'Has any relative died of heart problems before the age of sixty?') as a crude way of measuring genetic predisposition.

Currently in most countries there is a moratorium on genetic testing. Companies are not allowed to perform more accurate and direct genetic testing – or even ask more detailed questions – for fear the answers will be used unethically. It could stigmatise the individual, prevent them getting work or obtaining further insurance or medical cover. The other reason is fear that the information may be used for other, sinister purposes, and uncovering family secrets like nonpaternity. Understandably, for their part, the insurance companies fear that well-informed potential customers could work out their own risks from their family history and if needed obtain their own genetic profiles without telling the company – thus tilting the odds of getting a payout in their favour.

Why do females usually outlast males?

E was eighty and, although fit and alert for her age, was depressed, with no immediate family and no will to live any longer. She had had a rich and varied life, not without its misfortunes. She first married a soldier after a whirlwind romance during World War Two. He died in the Normandy landings in 1944. She was remarried three years later to a civilian who died in 1959 of tuberculosis. Her third husband was one year older than she was and healthy for most of his life, until he had a series of heart attacks and died three months later at the age of 79, leaving her a widow for the third time. She never got over being alone.

Why males of most species (including humans) die earlier than females has been a bit of a mystery. One hundred years ago, US men and women lived to around the same mean age – premature female deaths due to childbirth levelling out the averages. Although more males than females are born, by the age of fifty, the ratio is similar. But by age eighty, females outnumber males two to one and by a hundred the ratio is nine to one. Currently, US women live eight years, and British women six years, longer than men.

There are a number of theories to explain the imbalance. An excess of risk-taking behaviour is one, resulting in threefold greater numbers of murders and accidental death in males. Heart disease accounts for 50 per cent of male deaths before 65 and is about five times commoner than in women. The protective effects of natural oestrogens before the menopause in females may also explain part of this. Another explanation comes from the larger physical size of most males. We have previously discussed how height equates inversely to longevity. Human males are roughly 7 per cent larger than females, which roughly equates to the mortality difference. Male hormones are also important, and one of the few benefits of being a castrato, (other than singing) is that you get to live longer.

Testosterone, the male hormone needed for strength and aggression, further dampens down the immune system and influences a number of diseases. US males aged thirty to sixty years are twice as likely to develop infectious or parasitic diseases as females. This has been found to be true of most animal species. This increased susceptibility may partly explain why heart-disease rates are so much greater in males. One of the latest theories is that the disease is triggered by minor infections in the arteries, which become inflamed, leading to damage, clots and blockages (atheroma). Testosterone levels in males under the age of sixty fluctuate more than in females and can increase dramatically during times of stress.

Recent surveys found that males are 25 per cent more likely to have heart attacks at the time of a crucial and tense football match, such as the penalty shoot-out between Holland and Germany in 1996 and England versus Argentina in 1998. Two South Koreans died instantly during the closing stages of their team's unexpected victory over Italy in 2002. An excess of male heart attacks was recorded in Tel Aviv during the Scud missile attacks by Iraq in 1991. It seems to be predominantly males who get too excited or anxious – which could be due to high peaks in their blood testosterone levels. Women seem to possess mysterious control mechanisms that prevent their emotions becoming fatal.

As previously discussed, the only compensation for poor short-lived males is they are less likely to have overreactive immune systems and so get fewer autoimmune diseases of the kind that plague females. They are also less prone to common forms of arthritis, such as osteoarthritis, or dementia. Males aged under fifty are less than half as likely to seek medical attention for minor ailments than females, suggesting they have a better quality (but not quantity) of life. Surveys unfortunately suggest that in most countries males are actually less healthy – but exist in blissful ignorance. The situation may be changing: predictions are that if current trends of smoking and alcohol abuse continue in young women, who are outdoing young men, the sex difference could even out by the end of this century.

Lumps and bumps

V had just turned fifty. She was slightly overweight and not very fit. Looking at herself closely one day, she was concerned to see small varicose veins behind her left knee. Her legs had also begun to ache when she stood for a long time. She knew her mum had also suffered from her veins, but attributed this to working in a shop, and being on her feet all day. Her father had just been diagnosed as having Parkinson's disease, which meant he had become very stiff and couldn't stop shaking. V was worried what the future held for her – from her legs to the more serious problems her father suffered from.

Varicose veins affect about 20 per cent of women, with symptoms ranging from mild itching to large twisted veins and disabling ulcers. It is twice as common in women as in men. The causes are primarily genetic and our twin study showed a heritability of around 60 per cent. It is only slightly associated with occupation and childbirth. If your mother suffered from it, your risk is increased two- to threefold, giving you about a fifty-fifty chance. Varicose veins are due to a change in the elasticity of the walls of the veins that makes its valves malfunction, putting pressure on the rest of the system, leading to enlargement and twisting of many areas. There is likely to be a shared predisposition to haemorrhoids (piles).

Likely genes involved in vein diseases are those altering collagen, which is a major component of the lining of the vessels, and a gene (thrombomodulin) involved in blood clotting. Preventive measures that are likely to help reduce the severity of vein problems include losing weight; exercising for short periods frequently; avoiding long periods of standing; and using compression support tights. V, unfortunately, with her family history, doesn't have a lot to look forward to, but, if she got herself a bit thinner and fitter, many of the diseases she is likely to be susceptible to will be milder and easier to cope with.

We have discussed how most well-known diseases have an important genetic component. A few disorders appear to be rare exceptions to this rule and appear to have virtually no genetic or heritable basis – like the common form of late-onset Parkinson's disease, which causes tremor and rigidity and in some cases dementia. This may be due to purely random events (luck) or to trauma, as the disease is much commoner in boxers, including Muhammad Ali. Another group of apparently 'nongenetic' diseases in which twin studies have been unable to find a significant heritability are mild intestinal disorders such as irritable-bowel syndrome (IBS), and 'work-burnout' disorder. These disorders may well be true examples of totally random or environmental causes, but are so unusual that it suggests there is some other unknown factor about these diseases, or the way they are currently classified, that eludes us.

9

Beliefs, Morals and the Afterlife

There is more to human life than just a mixture of our genes and environment. The rich variation in our thoughts and beliefs and our preoccupation with philosophy, good and evil and the afterlife mark us out as distinctly human. These cultural traits are all assumed to be a result of the teachings passed down to us by the complex social network and civilisation we have created – but are they?

Political leanings

H came from a long heritage of forefathers with strong right-wing views and conservative values. He was taught to be polite to women, but that their place was in the home. He was opposed to socialism and taxes, which in his view encouraged laziness. He taught his 21-year-old son C to behave as he did and believed he shared his values. When C wanted to leave to study in a large liberal-minded city, his father was worried whether the family's influence would wane and his son's attitudes would be altered (for the 'worse') by his fellow students. To what extent are our social attitudes or political beliefs ingrained or learned?

Political views clearly run in families – one only has to look at the Kennedy or Bush dynasties to see this. However, much of this may be due to environmental factors. In the US, wealth and power are an essential attribute for a political family, suggesting the environment is crucial. So are our views all shaped by our cultural surroundings, the views of our parents and life experiences? Twin studies in the US and UK have looked at this question in large studies totalling thirty thousand twins and their families. They showed a clear genetic component to a number of attitudes and political views, including how firmly individuals hold right-wing fundamental conservative views such as the role of women and the church in society, the right to defend yourself with firearms, and believing in compulsory military service. People of similar beliefs and attitudes tend to marry each other – presumably because this is one fewer thing to fight over. Strangely, we are not particularly attracted to partners of the same personality type (e.g. open, fussy, extrovert, jealous or anxious) as ourselves – so there are still plenty of other areas for argument and potential conflict.

A new study using our cohort of three thousand (mainly female) twins in the UK found that political preference had a strong genetic component. This was most clearly seen for the voting intentions of individuals either for or against the more right-wing Conservative Party, where we found a strong heritability of over 71 per cent, with only a small contribution of environment and upbringing. A lesser but still significant genetic influence was seen for the more middle-ground Liberal Democrats (42 per cent) and the most left-of-centre mainstream party, Labour (37 per cent), which showed a much greater effect of environment and upbringing.

In countries such as the US with a presidential system, the attractiveness of the leader may be an important factor. Most US presidents have been taller than their opposing candidates (with the exception of Richard Nixon). Many have been attractive. Computer tests have shown that Bill Clinton had one of the most symmetrical faces possible, which might have accounted for some of his amazing popularity even in the face of adversity. It's unclear whether Tony Blair has been tested, but it can't be long before would-be presidents and prime ministers get symmetrical plastic-surgery makeovers by their image consultants. Baldness also seems to be an adverse factor in popularity of leaders – a fact which the British Conservative Party should take note of.

H's son C, if he moves to the liberal environment of the city, is unlikely to alter dramatically in terms of his attitudes and convictions. His upbringing and inherited tendencies should prove fairly resistant to any effect from a more liberal environment. He should also be quite resistant to party-political advertising and propaganda, unless his personal circumstances change markedly – which is bad news for political party fundraisers.

Criminal behaviour

S was arrested after being found with a murder weapon in his hand, having shot a defenceless shopkeeper dead in an unprovoked frenzied attack. At his trial, his defence lawyers claimed that his condition was genetic. On researching the family history, they had found that individuals in the previous three generations of his family had committed acts of extreme violence, including murder, arson and rape. They claimed he had a rare genetic mutation, which failed to control the excess chemicals in his brain (dopamine) that provoked violent impulsiveness. They asked that his family be tested genetically and that he not be treated as a murderer.

Whether criminal behaviour has a genetic basis or not is highly controversial and has been the subject of debate for over a hundred years. One of the first studies clearly showing a genetic influence was a Danish adoption survey that found that rates of criminal behaviour were greater in adopted children *from* criminal parents than in adopted children *brought up* by criminal parents. There is now good evidence that certain genes may be crucial. Rare mutations in the

MAOA gene (which controls the enzyme of the same name) on the X (sex) chromosome increase the likelihood of criminal behaviour. The case scenario above is that of a defendant called Stephen Mobley in the USA. His legal defence team quoted research studies of some recorded rare families, such as a notable large violent family from Holland, of which at least eight males had gene defects in the MAOA gene as described in the scenario. These are real 'neighbours from hell' and should be given a wide berth, as they are highly prone to acts of mindless violence, including murder, arson and rape – often of their close relatives or work colleagues.

Brain research has now established the areas of the brain that are altered in these individuals. They are all in the frontal or prefrontal cortex, the area dealing with emotion. Everyone has aggressive emotions but most of us suppress or modify them when we receive signals from our environment that they are inappropriate or causing fear in others. Extremely aggressive or antisocial men have a low threshold at which emotion is triggered and an impaired mechanism for controlling it. The main brain chemical that is implicated is serotonin, and abnormal levels (high levels in blood and low levels in the brain) have been found in aggressive criminals and convicted children. Low levels or activity of the enzymes MAOA and tryptophan hydroxylase (TPH) reduce brain serotonin levels and consequently modify emotional responses to stress. Criminality and antisocial behaviour is often related to events in childhood. Maltreatment by violent parents is an important factor, increasing the risk of later criminality by 50 per cent, and more if the abuse occurs very early in life. Maltreatment has been itself shown to alter brain chemicals in children. However, most maltreated children don't end up as antisocial criminals.

The serotonin-regulating genes TPH and MAOA have been found to have mutated more often in violent aggressive criminals. A recent 23-year follow-up study of a thousand children in New Zealand has demonstrated that just having one of the mutated genes may not be sufficient in itself to lead to anti-social behaviour and criminal acts. Young men who had the abnormal MAOA gene that leads to low enzyme activity (and therefore low brain serotonin) were ten times as likely to commit violent crimes only if they had also been seriously maltreated by their parents (8 per cent of them were); 85 per cent of this genetically predisposed and abused group also showed antisocial behaviour. Those with the same at-risk gene mutation who came from stable backgrounds without abuse were at hardly any greater risk of violence, antisocial behaviour or criminality.

Violence and criminality are therefore not predestined, and can be modified by society. Experiments that deprive baby lab animals of affection show that the expression of certain genes (such as oxytocin receptors, important in bonding) may be permanently altered – and influence brain dopamine and serotonin levels. Using a genetic 'excuse' is unlikely to cut much ice with judges – even less than the mitigating effects of an uncaring society or poor education currently does – as you are even more likely to repeat the crime. Whatever the final list of the exact genes involved, we know from the fact that – since 98 per cent of burglaries and 90 per cent of murders are committed by males – the biggest risk factor is the male Y chromosome, which is not much better an excuse than saying society is to blame.

The darker side of the Y chromosome

U was looking forward to her first date with C. They'd met several times at work and she had been attracted to him, although he had a reputation for being a bit aloof and cold. After a pleasant meal with wine, they started flirting in a light-hearted manner. He offered to drive her home and she accepted. She pecked him on the cheek as she left the car and said thank you. He insisted on 'a coffee'. She reluctantly agreed but made it plain that that was all that was on offer. As soon as they got inside, he kissed her passionately and started to undress her. She resisted and tore herself away, telling him to leave. He pushed her on to the sofa and with more force tried to remove her skirt, saying he knew she wanted him now. He was very strong and she was unable to move. He started unzipping himself and, when she screamed loudly, he slapped her but she kept screaming until her next-door neighbour knocked on the wall. C got off her and ran to the door telling her to calm down and forget what had happened.

Rape is very common, the frequency depending on how you define it. When rape is loosely defined as forcing a person to have sex against her wishes (whether this involves physical violence or not), it is reported by one in six women and has been estimated to affect as many as one woman in four (as many women may not report rape). Rates are even higher in some groups, such as 35 per cent in US female military recruits and in South Africa, where about one person in eight is infected with HIV. This is one of the highest rates in the world, with an estimated 2 million women and children raped per year. Surveys in the UK and US suggest that 4–6 per cent of women report having been raped with physical force and violence at some time.

Contrary to common belief, nearly two-thirds of rapes are by a close acquaintance – partners, husbands, dates or lovers – and 14 per cent of married women state they have at some time been raped by their husbands. As rape is not usually committed by strangers, it may therefore be an unpleasant extreme pattern of male behaviour. There is some support for this. In a psychology study, male volunteers were presented with a hypothetical scenario where the subjects could have sex with an attractive but non-consenting woman, anonymously without violence, fear of discovery or loss of reputation. A third of men responded that they might do it. In tests, many men (despite not condoning it) find visual depictions of rape (without overt violence) to be stimulating, but not more so than consensual sex. In other words, they are programmed to be aroused by the sex act or the associated power and control in most situations or environments. Only in the context of physical violence or visible signals of distress in the female are normal men turned off.

Rape occurs in all societies and cultures and is unfortunately part of human history. In many primitive tribes, the principal aim of warfare was to obtain women forcibly from other tribes – which indirectly led to gene mixing. In modern times, rape unfortunately often still goes hand in hand with warfare (Bosnia, Congo), suggesting that the motivations for both have similar evolutionary origins. Dominating other tribes can be successfully effected by physical or genetic means. There are no published studies examining the genetics of being a rapist, but the personality profile of a traditional stereotype 'stalker' rapist is a man with low status in society and therefore the smallest chances of obtaining a consenting mate in normal circumstances. On average, rapists are believed to be young, of low income and low educational status, insecure and unattractive, although this probably reflects only the minority of rapists who get caught attacking strangers. Little is known about acquaintance rapists. Other studies have suggested that stranger or stalker rapists have above average reproductive success and they tend to target women who are in the most fertile age range. A recent highly controversial study found that conception rates of rape victims are twice as high (6 per cent) compared with those women having random consensual sex. If true, this suggests that rapists may subconsciously pick up signals of the timing of their victim's maximum fertility.

Examples of rape with an unwilling or unreceptive partner in our close relatives the apes are fairly rare – an exception being adolescent and rather puny male orang-utans. Many other animals, including lobsters, fish, turtles, cats and bats, are known to rape their females – often with violence. Minks can apparently ovulate only after violent sex – presumably for males to physically prove their worth. Some males, such as sagebrush crickets and scorpion flies, have evolved clamps to keep a reluctant female in place, if they can't be tempted with food. Rape, therefore, may have evolved as a survival strategy in our male ancestors to combat the fact that a few males tend to be very successful in the mating business, at the expense of many other males who fail to find a mate.

This unequal playing field is probably still true for modern males: in most countries, due to oversuccessful older males, there are usually more young unattached men than unattached women. One desperate way out of this genetic oblivion, if you were not of high enough status, was sadly to become a hit-and-run rapist. Although all of us (men and women) unfortunately have ancestors who were rapists (and therefore have their genes), in normal circumstances, the vast majority of men are not rapists. In most cultures it is severely punished and the perpetrators stigmatised. When these restraints are relaxed, as in times of war, rape becomes more common. Whether rapists predominantly have out-of-control sexual urges or are driven by a need for power and control over women (or a mixture of both) is unclear. The phenomenon of male rape is undoubtedly more about power rather than sexual orientation or reproduction – and studies of sex drives and testosterone levels suggest both factors are important. As with criminality, it takes more than just an inherited predisposition to become a rapist and a series of environmental events such as childhood abuse are likely to be necessary. In a study of male US navy recruits, those admitting (before joining the service) to having raped (an amazing 11 per cent) had a high chance of having been physically or sexually abused themselves.

For women, though, the intense emotional trauma of rape is often lifelong, greatest in younger women (the most fertile) and probably related to the drives that make us all want to maintain control and choice of our genetic line. As most rape is committed by someone known to the victim, as with other forms of violence or bullying, it is quite possible that there is an associated common personality profile of a rape victim. Unpublished twin studies have apparently found that being a rape victim was also a heritable trait. Quite what this heritable victim factor is is unclear. Studies of female navy recruits who admitted to coercive sex showed that those who had been sexually abused (8 per cent) had much higher chances of later being raped. Studies in UK women also showed the risk of rape to be increased threefold in childhood abuse victims and was also associated with a high risk of domestic violence. Whether or not some victims are genetically or environmentally predisposed, it is not to their advantage. Most females of a variety of species risk injury or death to avoid a rapist and, if impregnated, produce less healthy children than if they get to select the father.

Skin deep

K always believed he was black. He never knew his father, whom his mother never discussed. He remembered his early school years with fondness, until the age of nine, when he changed to a new and larger school. He wanted to mix with the other black kids, but was shocked when they told him he wasn't really black and couldn't be one of them. He came back to his mum in tears. She explained to him about his (white) English father and how, because K was a mixture of two races, his skin and hair were lighter. She told him that as far as she was concerned he was a perfect black kid – and shouldn't take any notice. He began to hate his white blood and his father. Much later, having eventually been accepted by his darker-skinned schoolmates, he became a professional soccer player. On one memorable away game he was taking a corner kick when the home crowd began heckling him and then, making ape noises, threw bananas at him. He and his teammates left the field for fifteen minutes – but the experience affected him deeply.

Why do people react so badly and base so many judgements on the basis of somebody's skin colour or perceived race? For several centuries Europeans held the view that they were genetically superior and more mentally developed than other identifiable races, such as Africans, Aborigines and Asians. They equated power, resources and literature with genetic superiority and treated other groups as inferior species – often inventing stories that any sexual union would result in deformed monsters, as the genes were incompatible. Until recently it was claimed that IQ tests revealed differences between groups based on skin colour – but these differences disappeared when you adjusted for social deprivation, cultural differences, education and health. There are count-less examples from around the globe of racism in the guise of superiority in some physical trait or another – whether Zulu against Bushmen in southern Africa, Japanese against Korean or caste racism in India. They are all useful excuses for claiming the other group are not of the same status or even species – and therefore justifying treating them as subhuman.

In southwest Africa, at the end of the nineteenth century, European hunting parties would record the number of animals killed and, alongside the gazelles and lions, proudly included tallies of the number of female native Bushmen shot. In the island of Tasmania off the south coast of Australia in 1830, a group of hunters decided to have a coordinated effort to get rid of dangerous animals that had been on the island for 7,500 years. After years of using them for target practice and dogfood, in this one massive island-wide hunt (called the black line) they cornered their prey. Just sixteen Aborigines remained and their bones were fought over by museums in Europe.

It is only fairly recently that genetics has confirmed beyond doubt that, wherever we live or whatever we look like, we are all descended from a small group of East Africans. As we have already discussed, study of our mitochondrial DNA (the type passed down the female line) has shown that we all share genes in common with an African Eve some 150,000 years, or around 7,500 generations ago. This original common ancestor gave rise to around thirty female clans that spread round the world. Over the years some movement has been in the other direction. Most modern Africans have some ancestors from Asia and Europe and Native Americans came from Asia. Our genes are continually mixing, despite attempts by racial purists to prevent this, which only temporarily halt the process.

Although you can test an individual and predict with reasonable accuracy what part of the world their ancestors lived in recently, there is no clear way of distinguishing races any more. For geneticists this word 'race' is not definable. We are one large African species that recently spread around the globe. Several groups got separated from each other and, because the groups were small, interbreeding took place, which exaggerated any odd mutations such as red or blond hair and big noses in northern Europeans, or shape of eyes and eyelids in Asians. Some of these mutations may have had advantages, such as shielding the eyes from bright sun or pale skin allowing better absorption of vitamin D; others probably developed just because a few females liked the look of it and it caught on. For most of human existence the groups grew apart from each other and had little contact, as there were so few people. The last twenty generations have seen a major reversal of this, with huge movements of people and consequently remixing of the genes. Despite our apparent diversity and the spread of *Homo sapiens* across the planet, we humans share many more physical similarities than most other mammal species, including apes.

Why does the amount of the pigment (melanin) in our skins and the shape of our eyes cause so many problems for humans? We don't call all people with ginger hair or without earlobes separate racial groups. Variations in only a handful of genes are responsible for skin and eye traits – a tiny proportion of the total genetic variation. Within any one 'so-called' racial group such as Japanese or Bantu there is five times as much variety in the individual genes of people in that group than on average you find between the groups as a whole. Let us use the analogy of a card game, and we'll call it Genealogy. If an English pack and a Turkish pack were randomly dealt to four native players in each country, there would be much more variety in the strength of each player's hand than the subtle differences between Turkish and English playing cards. Despite all our migrations and mutations, 93 per cent of all human genetic variation and mutations can still be found in a few African tribes, closest to our common ancestors 150,000 years ago.

It has been calculated that 80 per cent of people living in one large geographical area today are descendents of any individual who reproduced successfully eight hundred years ago. Thus if an African emissary settled at an English court in the 1400s, or Chinese traders visited Africa, or a Venetian trader settled in Asia, everyone in those areas today would have eventually inherited some of these genes. Nowadays, someone walking in the street in New York cannot easily distinguish passers-by as potential distant relatives from Ireland, Denmark, Holland or the Ukraine. However, if they had Asian facial features or had black skin they would automatically assume they were from a 'different tribe'. We've seen how this assumption is usually wrong – with half of black Americans having recent white ancestors and many whites having recent black ancestors (such as some of President Jefferson's descendents). Ethiopians have many more genes in common with Europeans than Bantus from South Africa. Koreans are similar to Japanese and Palestinians are virtually indistinguishable genetically from Israeli Jews (confirming their common ancestors). Whether this will one day help or hinder racial harmony is another matter.

Our ancestors for most of our evolutionary period lived in wandering bands of twenty to fifty people, sometimes coming together into tribes of at most two hundred people. Everyone would know everyone else's name and be related either by marriage or by genes. Within this cosy circle, people must have felt reasonably secure. Once outside it, things would be very different – suspicion and wariness were important for survival. When, five thousand years ago, people started to get together in bigger collaborative groups, there was still a need for small-group identity and a circle of people you knew and trusted. Leaders saw that having discrete tribes would be counterproductive to the group and got together with priests to give people common identities via religions, citizenship, uniforms and a feeling of common purpose. Despite this, people still clung to their original perceptions of their roots and tended to form smaller ethnic groups with common cultures that they could identify with, just as they do today in many countries with a belief in common ancestry.

We've seen how racism is often a convenient cultural excuse for wrongdoing – but is it hard-wired into us or a cultural survival tool to pick allies and enemies handed down through the generations? In other words, is racist behaviour genetically determined? Twin studies were performed over twenty years ago that seem to have escaped public attention – and nowadays probably wouldn't have survived political correctness. The first twin study was conducted in the mid-seventies in England on 825 pairs and their spouses. Twins scored their feelings on a five-point scale for their agreement or disagreement with statements such as 'coloureds are innately inferior', 'we should make discrimination illegal' and 'we should encourage racial segregation'. The results were consistent and quite disturbing: the scores showed a wide (and frank) range of opinions. Fifty to sixty per cent of the variation between people was due to genes – with cultural factors accounting for only about 20 per cent. A similar and larger survey on 3,810 Australian twins was repeated in the 1980s and the results were consistent, showing a clear genetic influence on attitudes to racism, such as apartheid (heritability 43 per cent) and white superiority (49 per cent).

Saying that we have inherent racist tendencies doesn't mean we should condone or ignore them. Legal and cultural deterrents will have an effect, as the risks of giving in to primitive instincts are outweighed by the benefits of modifying behaviour. As world travel increases and the last isolated populations start mixing again, the physical and genetic differences between us will become ever more blurred. Some scientists predict that we will eventually all have the very light-brown skin colour. Racial tensions and obvious differences in physical characteristics and the stereotypes they produce may therefore not be an issue for ever. Clever psychology studies have shown that racial prejudices and stereotypes can be greatly diminished when 'racial' groupings are replaced by alternative coalitions or groupings. Before we get too excited about racism disappearing, let us look to recent lessons learned from fights between physically and genetically indistinguishable groups such as Serbs and Croats and Indian Muslims and Hindus, which teach us the sobering fact that, even without race as an excuse, humans will undoubtedly invent something else to fight each other over.

Genes for God?

S came from a good religious family of East European Jewish origin going back several centuries. Her father, H, was a businessman, and his uncle had been a rabbi. Her mother was also Jewish and was the more strictly religious and spiritual of the two. When H's wife died unexpectedly age fifty, he confided to his three daughters that he now had doubts about the existence of God. S later married (a Jewish but 'non-religious' boy) and had children of her own, but she never lost her strong religious beliefs and became the only practising member of the family. Her other sisters admitted also to having become agnostics like their father. She couldn't think of any particular event or experience that had made her different from her sisters – but maybe there was something she'd forgotten.

Religion and spirituality are commonly described as the essence of being human. Religious beliefs were for a long time viewed as evidence of superior human cultures and the role of societal and parental influence was seen as an essential part of this. There are many different components to being religious, which makes it difficult to define. These include religious identity or affiliation (calling yourself Jewish or Christian), religious practice (going to church regularly) and religious beliefs, which commonly include a concept of spirituality.

Twin studies have looked in detail at the question of religion by using diverse cultural and ethnic groups, including both European and African-Americans in the US, as well as subjects from Holland, Finland and Australia. They have all shown clearly that religious affiliation is totally cultural, with people preferentially choosing spouses of the same religious group. That was not surprising. However, in all countries and groups there was a genetic influence on both religious beliefs and actual practices, with heritability estimates averaging around 40 per cent. There was still an important effect of upbringing and family environment, which was reassuring. A US study of twins reared apart and raised in different religious back-grounds confirmed these findings. Psychological tests have shown that being genetically susceptible to religious beliefs doesn't automatically confer selfless or altruistic behaviour, suggesting that this needs to be taught to all children.

Religious and political beliefs depend to some extent on willingness to conform, which is itself a genetic trait. Conformity was probably an important part of our ancestors' tribal life and crucial to our human social networks. As Voltaire said, 'If God did not exist, it would be necessary to invent him.' Most people form judgments or opinions based on a mixture of their own views and those of other people around them. It would be foolish to ignore the opinions of hundreds of other members of your tribe or family in all situations, as they may be correct and you may actually be wrong. People who went 'with the flow' and followed the majority opinion may have been generally favoured and survived longer – allowing 'conformity' genes to flourish. Studies have shown that most people will change their opinion to conform with the overwhelming majority view.

Clever experiments were performed in which a dozen stooges stated (falsely) in turn that one particular straw was the longest in a bunch. On hearing this from twelve 'independent' witnesses, the majority of people would give the same demonstrably false answer. There are many examples from history of an overwhelming tendency to conform, such as in Nazi Germany or Maoist China. A small proportion of non-conformists might flourish in the system, if their numbers were small and they were tolerated by the majority, since, by not obeying all the tribal or religious rules, they may have been able to reproduce more or avoid death.

Morality or some form of preprogramming about right and wrong is generally believed to be in-built in humans and our willingness to accept the will of the community is a defining part of being human. Whether this was given to us by divine intervention or is just another evolutionary gift is for others to determine. Religion and group altruism works well only if 'cheats' are kept in check by fear of retribution by super-natural forces. Organised religion and group beliefs (and conformity) have a lot going for them in evolutionary terms. These beliefs may have started in some form as long as 150,000 years ago, when modern humans and language first appeared. Since then the tendency has helped tribal morale, improved efficiency and reduced tribal disputes, improved fighting abilities and united the tribe against common causes of 'nonbelievers'. In a more general sense it has helped humans cooperate for common good to improve survival and reproduction. In the modern world, twin studies have shown that a religious upbringing reduces the risk that children will abuse alcohol and drugs, and later develop depression, and has a generally beneficial effect on health. The downside is that it is associated with prejudice and intolerance of others and, although it may restrain inherent violent tendencies to some degree, this may protect only those who share the same beliefs, and often leads to wars.

Most of us have looked to the 400 billion other stars in the sky in our galaxy and thought of the 100 billion other galaxies and marvelled at our universe. Some individuals have pondered the chances that there would be beings like ourselves on another planet thinking about the universe. The atheist Stephen Hawking put these chances into perspective when he asked why the universe is so close to the dividing line between collapsing again and expanding indefinitely. He estimated that if the rate of expansion one second after the Big Bang had been less by one part in 10^{10} (i.e. very small), the universe would have collapsed after a few million years. If it had been greater by one part in 10^{10}, the universe would have been essentially empty after a few million years. In neither case would it have lasted long enough for life (and us) to develop.

When you add to these odds, the various fortuitous comet and meteorite incidents, the cooling of the earth to produce vapour, water and atmosphere and the combining of chemicals to produce life in the primordial soup, it's not hard to see why most people believe in a divine architect. Non-believers have either to accept that we've been uncommonly lucky and rather alone or to find some physical explanation of why the universe is the way it is. One such explanation is that our universe is only one of multiple universes, all with different laws of nature. Those universes with the right set of conditions most likely to give rise to life occasionally suc-ceed. Life in other universes may involve a replicating machine similar to DNA, which evolves to become more complex. Eventually over millions of years evolution pro-duces beings with brains big enough with billions of neurons acting together in a self-organised manner to achieve consciousness and to conceive of such concepts as God and cosmology – and to ask such questions as Why?

Whether you believe in the divine architect, the lucky break or being part of infinity, we should all believe in some form of life after death – either through our souls in heaven or through our genes on earth.

10

Conclusion

In the previous chapters we have discussed how genes can influence virtually all aspects of our lives, and as a consequence some readers may be depressed and others may be relieved. Yet there is still a significant role for conscious decision making in altering our behaviour and environment. Understanding our genes enables us to explain why humans are so diverse and unpredictable.

We've seen how, by exploring life from our conception to our death and all the events in between, genes influence our lives and behaviour. But this might suggest we are just slaves to our genes. This idea is abhorrent to many people and implies a rigid genetic determinism and purely instinctive behaviour, leaving no room for environment or culture. We know this is wrong and have seen how every trait, even the most genetic, requires other factors, such as environment, particular behaviour or other genes, to manifest itself

Redressing the balance (ironing out your genes)

M was 55 years old and had just finished caring for her elderly parents, who had recently died. She wanted to concentrate on her own wellbeing and was planning an enjoyable retirement in a few years' time. She had health insurance and went for a full medical check-up. This revealed she was slightly overweight with a BMI of 27; her blood pressure was normal but her blood cholesterol level was slightly raised. She told the physician of her family history of heart disease and dementia and he recommended a screening battery of fifty genetic tests. The results came back a week later and showed her to have the high-risk form of the ApoE4 gene; the other lipid (fat) genes were normal, as were her heart ACE genes. She had an unhealthy form of the leptin gene, which meant carbohydrates were more rapidly converted to sugar, but efficient liver metabolism genes, which allowed her to drink alcohol without problems. On her doctor's advice she went on an exercise regime, swimming twice a week and brisk walking for twenty minutes four times a week. She lost 10 kilograms and altered her diet to include more fruit and vegetables, lentils and nuts, less red meat and fewer dairy products – and a glass of red wine every evening. She stopped eating carbohydrates completely. She enjoyed her retirement and remained free of major health problems.

Obviously this scenario is too good to be true. Currently DNA testing for common diseases (heart disease, arthritis etc.) is too crude to provide a sufficient degree of certainty to make clinical or major lifestyle decisions. At the moment the most accurate and useful genetic test available for common diseases costs nothing and can be tested anywhere in the world – it's called a family history. The scenario does, however, give us a glimpse of the future of genetic testing, and how we can change the direction our genes might otherwise take us. Having seen how crucial genes are on the events and course of our lives, it's natural to ask what we can do to optimise our genes and environment and deal with any problems.

The first step is to try to unravel your family history. Where possible, draw your own family tree for posterity. Ask your elderly relatives about your family secrets before they dis-appear with them. When you do have medical problems, and need help, make sure you tell people such as your doctor your full family history – so the professional advice you receive is based on all the facts.

If you have children or grandchildren or are thinking of having them, it's important to realise that genetic instincts inherent in you and your child will naturally guide you through most of the ups and downs. Obviously, the environment that children are brought up in does matter. However, the fact that brothers and sisters in the same family can vary so much implies that our ability to change them solely via the environment is limited, compared with the power of the genes. Let us return to the analogy of the swimming pool used in Chapter 1. Most families keep the water depth (environment) fairly constant at average levels – it would have to change a lot to make any difference to the risk that a genetically predisposed child might 'drown'. Some families may naturally keep their 'environmental' pools full to the brim or nearly empty – and changing the level in these cases can make a big difference. A simple example is tooth decay: even if children are genetically predisposed, optimising the environment with perfect dental hygiene and avoiding sugar will totally prevent the problem. Similarly, to see whether individuals have genetic potential in, say, sport, music, or art, they need a minimum amount of environmental stimulation to realise any talent.

As for our human urges, addictions and cravings, admitting to yourself that you have innate genetic impulses is the first crucial step to dealing with life's problems. Everyone has a different set of drives and strengths and weaknesses. Recognising the various forms they take is important so you can focus on alternative and less damaging reward strategies that your body and brain need to keep themselves happy – without harming or killing yourself too early.

Dangerous cravings can be controlled purely by old-fashioned human willpower – an ability to resist natural urges that no other animal possesses. The importance of willpower was demonstrated by a study performed on four-year-old children who were left alone in a room with a marshmallow in front of them, and told that, if they didn't eat the tempting item for fifteen minutes, they would be later rewarded and allowed to eat two. Nearly all the kids ate the marshmallow. However, researchers found that, ten years later, those children who had waited longest were the most successful socially and academically. Willpower is only a mildly genetic trait, and therefore something we can all improve on. Alternative strategies for those that lack the iron discipline are to transfer to safer addictions that still produce brain rewards – such as substituting methadone for heroin, nicotine patches for cigarettes and competitive sports for gambling. In the future there will be many safer chemical alternatives for a variety of addictions, such as the latest dopamine-blocking drugs (Dopomine D3 receptor antagonists) that reduce the rewards and addictive potential of cocaine users (albeit for the moment in rats).

The modern perception is that genes cause diseases. Every day we hear of the discovery of another disease gene. From the individual gene's point of view, making your human host diseased is a bad idea and often a genetic dead end. Our genes, if they had thoughts, would want us kept alive, attractive and fertile for as long as possible so they can replicate. So, it's clear that genes don't cause disease: diseases are merely a side effect of some unforeseen interaction between a gene and an environment when the normal defence and survival mechanisms don't quite work optimally. We've seen many common examples of this, such as genes preventing infection, inadvertently causing auto-immune disease and allergy, or genes for maleness leading to heart disease. The rarer genetic diseases are often just due to unfortunate genetic mistakes from replicating errors or provide an advantage when in a single dose.

Everyone is likely to inherit predispositions to a disease or two, plus a few less desirable character traits. Rather than merely blaming your genes, spare a thought for them: they're just trying to do their job and make it through the next 5 million years or so, whereas we just have to get through three score and ten. To prevent most diseases, there's no quick or easy solution. You can cross your fingers and go with the flow, or you can find out where your main risks lie and have a long-term goal of adapting your lifestyle whenever you can. Living more like our ancestors isn't a bad general plan – although try to pick the nonviolent aspects of prehistory and avoid club fighting. Regular exercise, even if only fifteen to twenty minutes per day, is important to keep your body fit and keep your metabolic rate up.

Diet is the other important aspect of our lives we can (in theory) control. In the 1990s we were encouraged to replace fats with carbohydrates and eat more fruit and vegetables. More recent evidence suggests that we are eating far too much carbohydrate and some fats, rather than being harmful, are actually good for us. It has been shown that unsaturated fats are beneficial in reducing heart disease and raise levels of protective fats in our bodies (such as HDL). We should therefore eat more 'ancient' foods such as vegetables, olive oil, fish and nuts (even peanuts are good!). At the same time we should reduce modern high-starch foods such as bread, pasta, potatoes and rice (unless it's wholegrain), as these cause rapid and unnatural rises in our blood sugar level, which overstimulates leptin and insulin levels. Eating healthier foods such as fruit or nuts before the temptation of a high-fat restaurant meal is one strategy to deal with fat and calorie cravings.

Finally, for a happy emotional and love life, don't be too depressed by all the statistics about the ideal partners people supposedly want. Whatever your size, shape, colour or age, surveys have found that the universal preference of men and women of all cultures is for kindness and understanding – a trait we can all improve on, regardless of our genes.

Variety is the spice of life

O was born to poor black parents in a deprived part of the country. She had an unhappy childhood, raised between her grandmother's and separated parents' homes. She was sexually abused by her relative's boyfriend from the age of ten and had a stillborn child at fourteen. She was bullied at school

*for being different. Although initially slim, she started over-
eating and her weight fluctuated wildly throughout her life.
She also had temporary problems with alcohol and drug use.
O, however, did have some innate talents. At nineteen she
became the country's youngest newscaster and then went on
to do more TV and had her own talk show. She made several
films and was nominated for an Oscar. Through a series of
successful media companies, she has become one of the
richest and most powerful women in the US. She has openly
faced her own problems and urges and helped herself and
millions of others to deal with them.*

The Oprah Winfrey story is obviously a unique one – given
the amount of success she has achieved. Yet the world is full
of millions of individuals like her – less visible people, who
have come from humble backgrounds and inauspicious
genes, overcome great internal and external obstacles and
transform their lives.

Studies of human behaviour based purely on evolutionary
biology can give the impression that we all act with the same
basic human instincts that we either evolved ourselves or
shared with other animals. If this were true, we would all be
much more similar than in fact we are. Why are we not all
aggressive, powerful, beautiful, fast-running, sex-obsessed,
cunning and selfish individuals? We all think we know
people like that (not ourselves, naturally), but there are many
more exceptions.

For every aggressive person, there's a meek one; for every
muscular one, there's a puny one; for every bright one, there's
a dim one; for every selfish human, there's a good, altruistic
one; and so on. Why do we find so much variety between any
two people and why is our appearance and behaviour much
less predictable that you might imagine? Is this the key to the
success of humans? It is likely that we have unpredictability
genetically hard-wired into us. Our predictable ancestors may
have all died out as they lost their lives to predators or other
humans in crucial situations. This may have occurred to those

predictable humans who always wanted to eat at the same time of day or, when faced with a predator of a certain size, always ran away, ran at the same speed and in the same direction and were easily eaten. Unpredictability and the ability to do the unexpected, such as occasionally throwing a stone, shouting, running up a tree or forming groups, may have been an important survival strategy that we still carry today. Having a variety of human behaviours and unpredictable emotions may also have evolved to reduce cheating in individuals and increase group cooperation.

The best human strategy for both the species and its individuals is to have a balance of different genes, attributes and behaviours. Humans are the most social of animals, as they have learnt to cooperate with one another, which has brought great rewards. This is another reason why humans needed to have inherently flexible behaviours, so they can adapt their tactics dependent on their surroundings and what others are doing. Human instincts inherited over millions of years by transmission of genes have given us vital strategies to survive and reproduce. However, unlike most other animals, humans have more of a menu to choose from. Although in life-or-death situations we usually use the fixed 'instant survival' menu, we can sometimes go 'à la carte' and pick different strategies from a wider genetic repertoire. The ability to do this probably comes from our unique human feature – our large intellect.

Our genes, perhaps ten thousand of them, programme how and at what speed our brains develop, and determine the basic and relative size of the areas of creativity, motor skills and emotion, rather like the blueprint for components of a powerful computer. However, the machine that is created is self-learning, picking up knowledge from the environment, with trillions of chemical connections that dwarf the small numbers of blueprint genes. In this way our brains can choose the strategy from the menu that best suits the situation – ranging from the extremes of our primitive instincts of surviving and reproducing at all costs, to acts of suicidal heroism and deliberately leading a life of celibacy. The amazing variety between us, and our ability to choose whether or not to follow our instincts, makes us uniquely unpredictable and very human.

If we are all genetically programmed to have a desire for the same type of person and have a fairly universal cultural and genetic view of what attractiveness is, why don't we all look more alike? There is much more diversity in appearance between us all than one might expect compared with other species. This may be for a reason. While on average we find the same traits attractive, as individuals it may not be in our genetic interest to all go after the most attractive person we have a chance with. Being witty, musical, smart or artistic are also attractive traits in humans.

Careful studies of animals (quails and salmon) have found that, whereas all will in the right circumstances mate with a universally attractive partner, a minority will have an additional strategy. These females also have preferences for 'ugly' or 'wimpy' mates, making them different from the crowd – and potentially giving them an advantage. Subconsciously, they may be selecting a mate whose genes are very different from their own, resulting in some cases in potentially healthier children. Picking the ugly duckling may explain why humans have such diverse appearance and tastes and why we can usually find something 'attractive' in everyone.

Playing out your hand

We are all a unique mixture of our genes and our personal environment and we should feel proud of them – they've been around for millions of years, and hopefully will be around much longer. It is fifty years since DNA was discovered and it is estimated that – despite the recent advances of the human genome project and the sequencing of the human genome – we currently understand only about 1 per cent of what we need to know about how our genes work. We know virtually nothing about how each cell works and how genes alter its original function. With 99 per cent of genetic science waiting to be discovered, just imagine what knowledge will be available to future generations.

Even if the details and fine print of how human genes work are still sketchy, you can now use the information gleaned from this book to understand the overall plan of how you work. Uncover what your genes are trying to do and how you can steer your body and mind in the right direction, accounting for the instincts that may be pulling you in the wrong direction. The genetic cards your parents held are a good guide to what makes you tick, but, for many traits, the exclusive mixture of genes you inherit will always produce a few unexpected surprises. There is always the element of luck in the game of life – luckily, twin studies have shown that this is free of genetic influence. In other words, accept the unique genetic cards you've been dealt, and use human free will to play the game that best suits them. Fulfil the genetic potential of 'good' genes and try to cope with the less favourable ones the best you can. To ignore your genes is like playing poker without looking at your hand. Have a peek and enjoy playing the game.

Glossary

amygdala: an almond-sized part of the brain responsible for controlling primitive instincts such as fear and guilt.

chromosomes: 46 wavy structures in every cell with around four thousand genes on each one. They are arranged in 23 identical pairs except for the sex chromosomes, which either have two X chromosomes for females or one X and one Y for males.

DNA: deoxyribonucleic acid – the chemical that genes are made of. Consists of four chemicals (bases) arranged in a double helix, like a spiral ladder.

dopamine: a hormone that sends important signals to the brain, particularly influencing emotion and behaviour.

enzyme: a protein that is responsible for carrying out a chemical reaction.

evolution: the process by which, in the right environment, random small changes in genes can increase survival or reproduction in the next generation and so lead to small changes in physical form or behaviour.

genes: the blueprints for the body that produce proteins and can self-replicate. There are around forty thousand of them in every cell in the body.

genome: the full sequence of 3 billion DNA bases that make up all the genes in an individual. Every genome is unique (apart from twins).

Heritability: a measure of the relative amount of genetic influence on a trait or disease in a population. It is usually presented as a percentage or proportion of the total variability between people that is due to genes. Values range from 0 per cent to 100 per cent, with any remaining percentage being made up of environmental factors and errors in measurement.

In general, heritabilities above 60 per cent are considered high and those below 30 per cent low. Heritability can be measured from twin studies and more crudely from family studies. It should not be confused with heredity, which is a general term for things that are passed from one generation to another.

HLA: human leucocyte antigen genes. These genes on chromosome 6 are carried on the surface of all our cells as unique identifiers, like barcodes, and are important in recognising foreign intruders and fighting infections. They are the most diverse set of genes in our bodies so everyone (except identical twins) has a unique combination.

hormones: proteins that are the main chemical messengers of the body. They tell it what to do. Examples include oestrogen, growth hormone, insulin and cortisone.

hypothalamus: an area in the centre of the brain crucial for controlling emotion, greed, appetite, sleep and sexual behaviour.

mitochondrial DNA: a special type of DNA in mitochondria that supplies energy to the cells of the body. The DNA contains 37 genes and is passed along unmixed via the mother's egg to the next generation. DNA matches are used to explore female ancestry, as only females transmit it.

mutation: a change or alteration in a part of a gene. This can happen by chance in the copying process. Often it has no consequence but it can have positive or negative effects. If common it is called a polymorphism.

pheromones: chemicals that transmit messages from one animal to another by subliminal smells – often related to sex, food or aggression.

Pleistocene: a key time period in world history from 1.6 million years ago to 10,000 years ago. Most of the evolutionary changes seen in modern man occurred in the latter half of this epoch from 500,000 to 50,000 years ago.

receptor: a chemical structure that is designed to pick up and respond to a particular chemical – such as a lock to a particular key

serotonin: a hormone (also known as 5–HT) in the blood and brain that has important influences on mood and behaviour. Its level is altered by drugs such as Prozac.

trait: a particular characteristic of a living creature. Examples include height, colour, personality and blood sugar levels.

Y chromosome: the chromosome that contains the male gene and is transmitted unmixed by males. DNA matches are used to explore male ancestry and paternity.

References

The latest genetic news and reviews can be found in weekly or monthly journals such as *New Scientist, Scientific American, Science, Nature, Nature Genetics* and *American Journal of Human Genetics*.

Chapter 1: Who Are We?
Gegax T, Hayden, T (1998), 'A Mysterious Baby Mix-Up', *Newsweek* 132, 38–9.
Cavalli-Sforza, L (2001), *Genes, Peoples and Languages* (Penguin Books).
Collins, F S, Green, E D, Guttmacher, A E, Guyer, M S (2003), 'A Vision for the Future of Genomics Research', *Nature*, 24 April 2003, 935–47.
Dawkins, R (1976), *The Selfish Gene* (Oxford University Press).
Diamond, J (1998), *Guns, Germs and Steel: A Short History of Everybody for the last 13,000 Years* (Vintage).
Foster, E A, et al. (1998), 'Jefferson fathered slave's last child', *Nature* 396, 27–8.
Johnson, W, Krueger R F, Bouchard, T J, Jr, McGue, M (2002), 'The personalities of twins: just ordinary folks', *Twin Res* 5:125–31.
Jones, S (1993), *The Language of the Genes* (Flamingo).
Khoury, M J, McCabe, L L, McCabe, E R (2003), 'Population screening in the age of genomic medicine', *N Engl J Med* 348, 50–8.
Linacre, A (2003), 'The UK National DNA Database', *Lancet* 361, 1693.

MacGregor, A J, Snieder, H, Schork, N J, Spector, T D (2000), 'Twins: novel uses to study complex traits and genetic diseases', *Trends Genetics* 16 (3), 131–4.

McGuffin, P, Riley, B, Plomin, R (2001), 'Genomics and behavior: toward behavioral genomics', *Science* 291, 1232–49.

Olson, S (2002), *Mapping Human History* (Bloomsbury).

Ridley, M (2000), *Genome* (HarperCollins).

Ridley M (2003), *Nature Via Nurture* (Fourth Estate).

Sykes, B (2002), *The Seven Daughters of Eve* (Corgi).

US Government Human Genome Project website: http://www.ornl.gov/hgmis/

UK Wellcome Trust human genome website: http://www.wellcome.ac.uk/en/genome/

'Views on evolution survey (2003)', http://abcnews.go.com/sections/science/DailyNews/evolutionviews.

Watson, J D, Berry, A (2003), *DNA: The Secret of Life* (Heinemann).

Winston, R (2002), *Human Instinct* (Bantam Press).

Chapter 2: Genes, Worry and the Early Years

Archer, G S, et al. (2003), 'Hierarchical Phenotypic and Epigenetic Variation in Cloned Swine', *Biol Reprod*.

Arnestad, M, Opdal, S H, Musse, M A, Vege A, Rognum, T O (2002), 'Are substitutions in the first hypervariable region of the mitochondrial DNA displacement-loop in sudden infant death syndrome due to maternal inheritance?', *Acta Paediatrica* 91, 1060–4.

Cohen, D B (1999), *Stranger in the Nest: Do Parents Really Shape Their Child's Personality, Intelligence or Character?* (John Wiley & Sons).

Davis, D L, Gottlieb, M B, Stampnitzky J R (1998), 'Reduced ratio of male to female births in several industrial countries: a sentinel health indicator?', *Journal of American Medical Association* 279, 1018–23.

Editorial (2003), 'Goodbye Dolly … leader and friends?', *Lancet* 361 (9359), 711.

Grech, V, Savona-Ventura, C, Vassallo-Agius, P (2002), 'Research pointers: Unexplained differences in sex ratios at birth in Europe and North America', *British Medical Journal* 324, 1010–11.

Helle, S, Lumma, V, Jokela, J (2002), 'sons reduced maternal longevity in preindustrial humans', *Science* 296, 10 May, 1085.

Krous, H F, Nadeau, J M, Silva, P D, Byard, R W (2002), 'Infanticide: is its incidence among postneonatal infant deaths increasing?: an 18-year population-based analysis in California', *American Journal Forensic Medical Pathology* 23, 127–31.

Mayor, S (2003), 'Human cloning may be impossible', *British Medical Journal* 326, 838–838.

Opdal, S H, et al. (2002), 'Possible role of mtDNA mutations in sudden infant death', *Pediatric Neurolology* 27, 23–9.

Ratjen F, Doring, G (2003), 'Cystic fibrosis', *Lancet* 361, 681–9.

Round, J E, Deheragoda, M (2002), 'Sex – can you get it right?', *British Medical Journal* 325, 1446–7.

Tamimi, R M, et al. (2003), 'Average energy intake among pregnant women carrying a boy compared with a girl', *British Medical Journal* 326, 1245–6.

Chapter 3: The Early Years

Ahmadi, K R, Goldstein, DB (2002), 'Multifactorial diseases: asthma genetics point the way', *Current Biology* 12, R702–R704.

Alarcon, M, et al. (2002), 'Evidence for a language quantitative trait locus on chromosome 7q in multiplex autism families', *American Journal Human Genetics* 70, 60–71.

Arbour, N C, et al. (2000), 'TLR4 mutations are associated with endotoxin hyporesponsiveness in humans', *Nature Genetics* 25, 187–91.

Baron-Cohen, S (2002), 'The extreme male brain theory of autism', *Trends Cognitive Science* 6, 248–54.

Bond, J, et al. (2002), 'ASPM is a major determinant of cerebral cortical size', *Nature Genetics* 32, 316–20.

Bouchard, T J, Jr, Hur, Y M (1998), 'Genetic and environmental influences on the continuous scales of the Myers–Briggs Type Indicator: an analysis based on twins reared apart', *Journal on Personality* 66, 135–49.

Buchsbaum, G M, Chin, M, Glantz, C, Guzick, D (2002), 'Prevalence of urinary incontinence and associated risk factors in a cohort of nuns', *Obstetrics Gynecology* 100, 226–9.

Campbell A (2003), 'Gender and Violence', *New Scientist* (24 May), 50–1.

Campbell, A (2003), 'Gender and Violence', *New Scientist* (24 May), 50–1.

Chenn, A, Walsh, C A (2002), 'Regulation of cerebral cortical size by control of cell cycle exit in neural precursors', *Science* 297, 365–9.

Comings, D E (1997), 'Genetic aspects of childhood behavioral disorders', *Child Psychiatry and Human Development* 27, 139–50.

Djurhuus, J C, Rittig, S (2002), 'Nocturnal enuresis', *Current Opinion in Urology* 12, 317–20.

Drayna, D, et al. (2001), 'Genetic correlates of musical pitch recognition in humans', *Science* 291, 1969–72.

Erata, Y E, et al. (2002), 'Risk factors for pelvic surgery', *Archives Gynecology Obstetrics* 267, 14–18.

Felsenfeld, S, et al. (2000), 'A study of the genetic and environmental etiology of stuttering in a selected twin sample', *Behavioural Genetics* 30, 359–66.

Fox, P W, Hershberger, S L, Bouchard, T J, Jr (1996), 'Genetic and environmental contributions to the acquisition of a motor skill', *Nature* 384, 356–8.

Gilad, Y, et al. (2002), 'Evidence for positive selection and population structure at the human MAO-A gene', *Proceedings of the National Academy of Science USA* 99, 862–7.

Gray, P M, et al. (2002), 'Biology and music: The music of nature', *Science* 291, 52–4.

Hammond, C J, Snieder, H, Gilbert, C E, Spector, T D (2001), 'Genes and environment in refractive error: the twin eye study', *Investigative Ophthalmology of Visual Science* 42, 1232–6.

Hauser, M D, Chomsky, N, Fitch, W T (2002), 'The faculty of language: what is it, who has it, and how did it evolve?', *Science* 298, 1569–79.

Helmuth, L (2001), 'Neuroscience. Dyslexia: same brains, different languages', *Science* 291, 2064–5.

Janata P, et al. (2002), 'The cortical topography of tonal structures underlying Western music', *Science* 298, 2167–70.

Koeppen-Schomerus, G, Stevenson, J, Plomin, R. (2001), 'Genes and environment in asthma: a study of 4 year old twins', *Archive of Disease in Childhood* 85, 398–400.

Lykken, D T, McGue, M, Tellegen, A, Bouchard, T J, Jr (1992), 'Emergenesis: Genetic traits that may not run in families', *Journal of American Psychology* 47, 1565–77.

McClearn, G E, et al. (1997), 'Substantial genetic influence on cognitive abilities in twins 80 or more years old', *Science* 276, 1560–3.

Mutti, D O, et al. (2002), 'Parental myopia, near work, school achievement, and children's refractive error', *Investigative Ophthalmology of Visual Science* 43, 3633–40.

Paulesu, E, et al. (2001), 'Dyslexia: cultural diversity and biological unity', *Science* 291, 2165–7.

Priftanji, A, et al. (2001), 'Asthma and allergy in Albania and the UK', *Lancet* 358, 1426–7.

Rutherford, J, McGuffin, P, Katz, R J, Murray, R M (1992), 'Genetic influences on eating attitudes in a normal female twin population', *Psychological Medicine* 23, 425–36.

Samaras, T T, Elrick, H, Storms, L H (2003), 'Is height related to longevity?' *Life Sciences* 72, 1781–1802.

Strachan, D, Wong, H, Spector, T (2001), 'Concordance and interrelationship of atopic diseases and markers of allergic sensitization among adult female twins', *Journal of Allergy and Clinical Immunology* 108, 901–7.

Turkheimer, E, et al. (2003), 'Socioeconomic status modifies heritability of IQ in young children', *Psychological Science* (In Press).

Wolfgang, E, et al. (2002), 'Molecular evolution of FOXP2, a gene involved in speech and language', *Nature* 418, 869–72.

Xu, J, et al. (2002), 'Major recessive gene (s) with considerable residual polygenic effect regulating adult height: confirmation of genomewide scan results for chromosomes 6, 9, and 12', *American Journal of Human Genetics* 71, 646–50.

Chapter 4: Genes and the Terrible Teens

Bataille V, et al. (2002), 'The influence of genetics and environmental factors in the pathogenesis of acne: a twin study of acne in women', *Journal of Investigative Dermatology* 119, 1317–22.

Benjamin, J, et al. (1996), 'Population and familial association between the D4 dopamine receptor gene and measures of Novelty Seeking', *Nature Genetics* 12, 81–4.

Bulik, C M, et al. (2003), 'The relation between eating disorders and components of perfectionism', *American Journal of Psychiatry* 160, 366–8.

Cordain, L, et al. (2002), '*Acne vulgaris*: a disease of Western civilization', *Archives of Dermatology* 138, 1584–90.

Eley, T C, Lichtenstein, P, Stevenson, J (1999), 'Sex differences in the etiology of aggressive and nonaggressive antisocial behavior: results from two twin studies', *Child Development* 70, 155–68.

Enoch, M A, Harris, C R, Goldman, D (2001), 'Does a reduced sensitivity to bitter taste increase the risk of becoming nicotine addicted?', *Addictive Behaviour* 26, 399–404.

Figes, K (2002), *The Terrible Teens* (Penguin Viking).

Gorwood, P, et al. (2002), 'The 5-HT (2A) – 1438G/A

polymorphism in anorexia nervosa: a combined analysis of 316 trios from six European centres', *Molecular Psychiatry* 7, 90–4.

Kipman, A, et al. (2002), '5–HT (2A) gene promoter polymorphism as a modifying rather than a vulnerability factor in anorexia nervosa', *European Psychiatry* 17, 227–9.

Klissouras, V, et al. (2001), 'Genes and olympic performance: a co-twin study', *International Journal of Sports Medicine* 22, 250–5.

Klump, K L, Kaye, W H, Strober, M (2001), 'The evolving genetic foundations of eating disorders', *Psychiatric Clinics of North America* 24, 215–25.

Koopmansl J R, et al. (1999), 'The genetics of smoking initiation and quantity smoked in Dutch adolescent and young adult twins', *Behavioural Genetics* 29, 383–93.

Lanouette, C M, et al. (2002), 'Uncoupling protein 3 gene is associated with body composition changes with training in HERITAGE study', *Journal of Applied Physiology* 92, 1111–18.

Lynskey, M T, et al. (2003), 'Escalation of drug use in early-onset cannabis users vs co-twin controls', *Journal of American Medical Association* 289, 427–33.

McGue, M, Iacono, W G, Legrand, L N, Elkins, I (2001), 'Origins and consequences of age at first drink. II. Familial risk and heritability', *Alcohol Clinical and Experimental Research* 25, 1166–73.

Sargent, J D, et al. (2001), 'Effect of seeing tobacco use in films on trying smoking among adolescents: cross sectional study', *British Medical Journal* 323, 1394–7.

Wade, T D, et al. (2001), 'The influence of genetic and environmental factors in estimations of current body size, desired body size, and body dissatisfaction', *Twin Research* 4, 260–5.

Woods, D, et al. (2001), 'Elite swimmers and the D allele of the ACE I/D polymorphism', *Human Genetics* 108, 230–2.

Chapter 5: Genes, Attraction and Sex

Bailey, J M, Dunne, M P, Martin, N G (2000), 'Genetic and environmental influences on sexual orientation and its correlates in an Australian twin sample', *Journal of Personality and Social Psychology* 78, 524–36.

Bailey, J M, et al. (2000), 'Do individual differences in sociosexuality represent genetic or environmentally contingent strategies? Evidence from the Australian twin registry', *Journal of Personality and Social Psychology* 78, 537–45.

Baker, R R, Bellis, M A (1995), *Human Sperm Competition: Copulation, Masturbation, and Infidelity* (Chapman & Hall).

Bartels, A, Zeki, S (2000), 'The neural basis of romantic love', *Neuroreport* 11, 3829–34.

Buss, D M (1989), 'Sex differences in human mate preferences. Evolutionary hypotheses tested in 37 cultures', *Behavioural and Brain Sciences*, 1–49.

Buss, D M (1995), *The Evolution of Desire* (Basic Books).

Buss, D M (2001), *The Dangerous Passion: Why Jealousy Is Necessary in Love and Sex* (Bloomsbury).

Buss, D M (2000), 'Desires in human mating', *Annals of the New York Academy of Sciences* 907, 39–49.

Clark, R D, Hatfield, E (1989), 'Gender differences in receptivity to sexual offers', *Journal of Psychology and Human Sexuality* 2, 39–55.

D'Amati, G, et al. (2002), 'Type 5 phosphodiesterase expression in the human vagina', *Urology* 60, 191–5.

Fay, R E, Turner, C F, Klassen, A D, Gagnon, J H (1989), 'Prevalence and patterns of same-gender sexual contact among men', *Science* 243, 338–48.

Fisher, H E (1992), *Anatomy of Love: A Natural History of Mating, Marriage, and Why We Stray* (Ballantine Books).

Gallup, G G, et al. (2003), 'The human penis as a semen displacement device', *Evolution and Human Behaviour*, 24, 277–89.

Gangestad, S W, Buss, D M (1993), 'Pathogens Prevalence and human mate preferences', *Ethology and Sociobiology* 14, 89–96.

Hare, B, Brown, M, Williamson, C, Tomasello, M (2002), 'The domestication of social cognition in dogs', *Science* 298, 1634–6.

Hardy S (2000), *Mother Nature* (Vintage).

Hughes, S M, and Gallup, G G (2003), 'Sex differences in morphological predictors of sexual behaviour,' *Evolution and Human Behaviour*, 24, 173–8.

Hu, S, et al. (1995), 'Linkage between sexual orientation and chromosome Xq28 in males but not in females', *Native Genetics* 11, 248–56.

Insel, T R, Young, L J (2001), 'The neurobiology of attachment', *Nature Reviews Neuroscience* 2, 129–36.

Jockin, V, McGue, M, Lykken, D T (1996), 'Personality and divorce: a genetic analysis', *Journal of Personality and Social Psychology* 71, 288–99.

Johnson, A M, et al. (2001), 'Sexual behaviour in Britain: partnerships, practices, and HIV risk behaviours', *Lancet* 358, 1835–42.

Jones, S Y (2002), *Y: The Descent of Men* (Little, Brown).

Judson, O (2002), *Dr Tatania's Sex Advice to All Creation* (Chatto & Windus).

LeVay, S (1991), 'A difference in hypothalamic structure between heterosexual and homosexual men', *Science* 253, 1034–7.

Lykken, D T, Tellegen, A (1993), 'Is human mating adventitious or the result of lawful choice? A twin study of mate selection', *Journal of Personality and Social Psychology* 65, 56–68.

Menashe, I, Man, O, Lancet, D, Gilad Y (2003), 'Different noses for different people', *Nature Genetics* 34, 143–4.

Milinski, M, Wedekind, C (2001), 'Evidence for MHC-correlated perfume preferences in humans', *Behavioural Ecology* 12, 140–9.

Miller, G (1999), *The Mating Mind* (Vintage).

Penton-Voak, I S, et al. (1999), 'Menstrual cycle alters face preference', *Nature* 399, 741–2.

Pitkow, L J, et al. (2001), 'Facilitation of affiliation and pair-bond formation by vasopressin receptor gene transfer into the ventral forebrain of a monogamous vole', *Journal of Neuroscience* 21, 7392–6.

Sasse, G, Muller, H, Chakraborty, R, Ott, J (1994), 'Estimating the frequency of nonpaternity in Switzerland', *Human Hereditary* 44, 337–43.

Savic, I, Berglund, H, Gulyas, B, Roland, P (2001), 'Smelling of odorous sex hormone-like compounds causes sex-differentiated hypothalamic activations in humans', *Neuron* 31, 661–8.

Savolainen, P, et al. (2002), 'Genetic evidence for an East Asian origin of domestic dogs', *Science* 298 (5598), 1610–13.

Schmitt, O P, et al. (2003), 'Universal sex differences in the desire for sexual variety', *Journal of Personality and Social Psychology* 85, 85–104.

Sobel, N, Brown, W M (2001), 'The scented brain: pheromonal responses in humans', *Neuron* 31, 512–14.

Soler, C, et al. (2003), 'Facial attractiveness in men provides clues to semen quality', *Evolution and Human Behavior* 24, 199–207.

Thornhill, R, Gangestad, S W (1994), 'Human fluctuating asymmetry and sexual behavior', *Physiological Science* 5, 297–302.

Townsend, J M, Levy, G D (1990), 'Effects of potential partners' physical attractiveness and socioeconomic status on sexuality and partner selection', *Archives of Sexual Behaviour* 19, 149–64.

Voracek, M, Fisher, M (2002) 'Shapely centrefolds? Temporal change in body measures: trend analysis', *British Medical Journal* 325, 1447–8.

Wedekind, C, Furi, S (1997), 'Body odour preferences in men and women: do they aim for specific MHC combinations or simply heterozygosity?' *Proceedings of the Royal Society of London Series B: Biological Sciences* 264, 1471–9.

Wedekind, C, Seebeck, T, Bettens, F, Paepke, A J (1995), 'MHC-dependent mate preferences in humans',

Proceedings of the Royal Society of London Series B: Biological Sciences 260, 245.

Wellings, K, et al. (2001), 'Sexual behaviour in Britain: early heterosexual experience', *Lancet* 358, 1843–50.

Chapter 6: Grown-up Genes, Instincts and Risks

Alonso, W J, Schuck-Paim, C (2002), 'Sex-ratio conflicts, kin selection, and the evolution of altruism', *Proceedings of the National Academy of Sciences USA* 99, 6843–7.

Benowitz, N L, Perez-Stable, E J, Herrera, B, Jacob, P, III (2002), 'Slower metabolism and reduced intake of nicotine from cigarette smoking in Chinese-Americans', *Journal of National Cancer Institute* 94, 108–15.

Blum, K, et al. (2000), 'Reward deficiency syndrome: a biogenetic model for the diagnosis and treatment of impulsive, addictive, and compulsive behaviors', *Journal of Psychoactive Drugs* 32 (Suppl), i–112.

Burnham, T, Phelan, J (2001), *Mean Genes* (Simon & Schuster).

Flanagan, O (2003), 'The Colour of Happiness,' *New Scientist* (24 May), 44.

Hajnal, A, Norgren, R (2001), 'Accumbens dopamine mechanisms in sucrose intake', *Brain Research* 904, 76–84.

Hariri, A R, et al. (2002), 'Serotonin transporter genetic variation and the response of the human amygdala', *Science* 297, 400–3.

Hettema, J M, (2003), 'A twin study of the genetics of fear conditioning,' *Archive of General Psychiatry*, 60, 702–8.

Jablonski, N G, Chaplin, G (2003), 'Skin Deep', *Scientific American* 13, 72–9.

Jarvik, M E, et al. (2000), 'Nicotine blood levels and subjective craving for cigarettes', *Pharmacology Biochemistry and Behaviour* 66, 553–8.

Kendler, K S, Jacobson, K C, Myers, J, Prescott, C A (2002), 'Sex differences in genetic and environmental risk factors for irrational fears and phobias', *Psychological Medicine* 32, 209–17.

Kendler, K S, Jacobson, K C, Prescott, C A, Neale, M C (2003), 'Specificity of genetic and environmental risk factors for use and abuse/dependence of cannabis, cocaine, hallucinogens, sedatives, stimulants, and opiates in male twins', *American Journal of Psychiatry* 160, 687–95.

Kendler, K S, Prescott, C A (1999), 'Caffeine intake, Tolerance, and Withdrawal in Women: A Population-Based Twin Study', *American Journal of Psychiatry* 156, 223–8.

Kim, U K, et al. (2003), 'Positional cloning of the human quantitative trait locus underlying taste sensitivity to phenylthiocarbamide', *Science* 299, 1221–5.

Kirkwood, T B, Austad, S N (2000), 'Why do we age?', *Nature* 408, 233–8.

Knutson, B, Adams, C M, Fong, G W, Hommer D (2001), 'Anticipation of increasing monetary reward selectively recruits nucleus accumbens', *Journal of Neuroscience* 21, RC159.

Loh, E W, et al. (2000), 'Association analysis of the GABA (A) receptor subunit genes cluster on 5q33-34 and alcohol dependence in a Japanese population', *Molecular Psychiatry* 5, 301–7.

Luo, M, Fee, M S, Katz, L C (2003), 'Encoding pheromonal signals in the accessory olfactory bulb of behaving mice', *Science* 299, 1196–1201.

Lykken, D T, Tellegen, A (1996), 'Happiness is a Stochastic Phenomenon', *Physiological Science* 7.

Lykken, D T, Bouchard, T J, Jr, McGue, M, Tellegen, A (1993), 'Heritability of interests: a twin study', *Journal of Applied Psychology* 78, 649–61.

Lynskey, M T, et al. (2002), 'Genetic and environmental contributions to cannabis dependence in a national young adult twin sample', *Psychological Medicine* 32, 195–207.

McKinney, E F, et al. (2000), 'Association between polymorphisms in dopamine metabolic enzymes and tobacco consumption in smokers', *Pharmacogenetics* 10, 483–91.

Madden, P A, et al. (1999), 'The genetics of smoking persistence in men and women: a multicultural study', *Behav Genet* 29, 423–31.

Martindale, D (2003), 'Burgers on the brain', *New Scientist* 2003; 177, 26.

Milinski, M, Wedekind, C (1998), 'Working memory constrains human cooperation in the Prisoner's Dilemma', *Proc Natl Acad Sci USA* 1998; 95, 13755–8.

Mineka, S, Ohman, A (2002), 'Phobias and preparedness: the selective, automatic, and encapsulated nature of fear', *Biology Psychiatry* 52, 927–37.

Mufano, M R, et al. (2003), 'Genetic Polymorphisms and Personality in Healthy Adults: A systematic review and meta-analysis', *Molecular Psychiatry* 8, 471–84.

Myers, D G (2000), 'The funds, friends, and faith of happy people', *Journal of American Psychology* 55, 56–67.

Ridley, M (1996), *The Origins of Virtue* (Penguin).

Rushton, J P, et al. (1986), 'Altruism and aggression: the heritability of individual differences', *Journal of Personality and Social Psychology* 50, 1192–8.

Sabol, S Z, et al. (1999), 'A genetic association for cigarette smoking behavior', *Health Psychol* 18, 7–13.

Sherratt, T N, Roberts, G (1998), 'The evolution of generosity and choosiness in cooperative exchanges', *Journal of Theoretical Biology* 193, 167–77.

Sipe, J C, et al. (2002), 'A missense mutation in human fatty acid amide hydrolase associated with problem drug use', *Proceedings of the National Academy of Science USA*, 99, 8394–9.

Swan, G E, Carmelli, D, Cardon, L R (1997), 'Heavy consumption of cigarettes, alcohol and coffee in male twins', *Journal of Studies in Alcohol* 58, 182–90.

Thun, M J, Henley, S J, Calle, E E (2002), 'Tobacco use and cancer: an epidemiologic perspective for geneticists', *Oncogene* 21, 7307–25.

Tumpey, T M, et al. (2002), 'Existing antivirals are effective against influenza viruses with genes from the 1918 pandemic virus', *Proceedings of the National Academy of Science USA* 99, 13849–54.

Wedekind, C, Milinski, M (2000), 'Cooperation through image scoring in humans', *Science* 288, 850–2.

Whitfield, J B, et al. (2001), 'Variation in alcohol pharmacokinetics as a risk factor for alcohol dependence', *Alcohol Clinical and Experimental Research* 25, 1257–63.

Woo, J M, Yoon, K S, Yu, B H (2002), 'Catechol O-methyltransferase genetic polymorphism in panic disorder', *American Journal of Psychiatry* 159, 1785–7.

Yamagishi, T, et al. (2003), 'You can judge a book by its cover. Evidence that cheaters may look different from co-operators,' *Evolution and Human Behaviour* 24, 290–301.

Young, M (2001), 'Kind and considerate. Reporting on work of Barclay P et al.', *New Scientist* 2001 (18 June).

Chapter 7: Minor Body Irritations

Abel, L, Casanova, J L (2000), 'Genetic predisposition to clinical tuberculosis: bridging the gap between simple and complex inheritance', *American Journal of Human Genetics* 67, 274–7.

Adult Dental Health Survey: Oral Health in the UK (1998), Office of National Statistics.

Aithinson, K J, Gill, M (2002), 'Pharmacogenetics in the postgenomic era', in Plomin, R, Defries, J, Craig, I, McGuffin, P (eds), *Behavioural Genetics in the Postgenomic Era*, 335–42.

Budiansky, S (2002), 'Creatures of our own making', *Science* 298, 80–6.

Comuzzie, A G, Allison, D B (1998), 'The search for human obesity genes', *Science* 280, 1374–7.

Conry, J P, et al. (1993) 'Dental caries and treatment characteristics in human twins reared apart', *Archives of Oral Biology* 38, 937–43.

Crosby, A, et al. (2003), 'Relation between acute hypoxia and activation of coagulation in human beings', *Lancet* 361, 2207–8.

Eriksson, E, et al. (2002), 'Diagnosis and treatment of premenstrual dysphoria', *Journal of Clinical Psychiatry* 63 (Suppl), 16–23.

Fisher, F J (1968), 'A field survey of dental caries, periodontal disease and enamel defects in Tristan da Cunha', *British Dental Journal* 125, 447–53.

Grady-Weliky, T A (2003), 'Clinical practice: Premenstrual dysphoric disorder', *New England Journal of Medicine* 348, 433–8.

Greer, I A (1999), 'Thrombosis in pregnancy: maternal and fetal issues', *Lancet* 353 (9160), 1258–65.

Gura, T (2000), 'Obesity research: Tracing leptin's partners in regulating body weight', *Science* 287, 1738–41.

Heath, A C, Eaves, L J, Kirk, K M, Martin, N G (1998), 'Effects of lifestyle, personality, symptoms of anxiety and depression, and genetic predisposition on subjective sleep disturbance and sleep pattern', *Twin Research* 1, 176–88.

Heikkinen, T, Jarvinen, A (2003), 'The common cold', *Lancet* 361, 51–9.

Hoffman, R (2003), 'Male Androgenic Alopecia', *Clinical and Experimental Dermatology* 27, 373–82.

Jepson, A, et al. (2001), 'Genetic regulation of acquired immune responses to antigens of Mycobacterium tuberculosis: a study of twins in West Africa', *Infection and Immunity* 69, 3989–94.

Lake, R I, Thomas, S J, Martin, N G (1997), 'Genetic factors in the aetiology of mouth ulcers', *Genetic Epidemiology* 14, 17–33.

Lavin, J H, French, S J, Read, N W (1997), 'The effect of sucrose- and aspartame-sweetened drinks on energy intake, hunger and food choice of female, moderately restrained eaters', *Internal Journal of Obesity Related Metabolics Disorders* 21, 37–42.

Mallinson, A I, Longridge, N S (2002), 'Motion sickness and vestibular hypersensitivity', *Journal of Otolaryngology* 31, 381–5.

Manek, N J, Hart, D, Spector, T D, MacGregor, A J (2003),

'The association of body mass index and osteoarthritis of the knee joint: An examination of genetic and environmental influences', *Arthritis Rheum* 48, 1024–9.

Marchetti, M, Pistorio, A, Barosi, G (2000), 'Extended anticoagulation for prevention of recurrent venous thromboembolism in carriers of factor V Leiden – cost-effectiveness analysis', *Thrombosis and Haemostasis* 84, 752–7.

Markus, M M, et al. (1998), 'A gene for universal congenital alopecia maps to chromosome 8p21-22', *Americal Journal of Human Genetics* 62, 386–90.

Maryetic, S, Gazzda, C, Pegg, G G, Hill, R A (2002), 'Lepturi: a review of its peripheral actions and interactions,' *International Journal of Obesity* 26, 1407–1433.

Michalowicz, B S, et al. (2000), 'Evidence of a substantial genetic basis for risk of adult periodontitis', *Journal of Periodontology* 71, 1699–1707.

Nieters A, Brems S, Becker N (2001), 'Cross-sectional study on cytokine polymorphisms, cytokine production after T-cell stimulation and clinical parameters in a random sample of a German population', *Human Genetics* 108, 241–8.

Pedley, T A, Hirano M (2003), 'Is refractory epilepsy due to genetically determined resistance to antiepileptic drugs?', *N Engl J Med* 348 (15), 1480–2.

Philpott S, Weiser B, Tanwater P, Vermund S, et al. (2003), 'CC Chemokine receptor S genotype and susceptibility ot transmission of human immunodeficiency virus type 1 in women.' *Journal of Infectious Disease*, 187, 569–75.

Pi-Sunyer, X (2003), 'A clinical view of the obesity problem', *Science* 299, 859–60.

Rivkees, S A (2003), 'Time to wake-up to the individual variation in sleep needs', *Journal of Clinical Endocrinology and Metabolism* 88, 24–5.

Rossouw, M, et al. (2003), 'Association between tuberculosis and a polymorphic NFkappaB binding site in the interferon gamma gene', *Lancet* 361, 1871–2.

Samaras, K, et al. (1997), 'Independent genetic factors determine the amount and distribution of fat in women after the menopause', *Journal of Clinical Endocrinology and Metabolism* 82, 781–5.

Siddiqui, A, et al. (2003), 'Association of multidrug resistance in epilepsy with a polymorphism in the drug-transporter gene ABCB1', *New England Journal of Medicine* 348, 1442–8.

Sindrup, S H, Brosen, K (1995), 'The pharmacogenetics of codeine hypoalgesia', *Pharmacogenetics* 5, 335–46.

Stern, K, McClintock, M K (1998), 'Regulation of ovulation by human pheromones', *Nature* 392, 177–9.

Susol, E, et al. (2000), 'A two-stage, genome-wide screen for susceptibility loci in primary Raynaud's phenomenon', *Arthritis Rheum* 43, 1641–6.

Taheri, S, Zeitzer, J M, Mignot, E (2002), 'The role of hypocretins (orexins) in sleep regulation and narcolepsy', *Annual Review of Neuroscience* 25, 283–313.

Takeda, N, et al. (2001), 'Neural mechanisms of motion sickness', *Journal of Medical Investigation* 48, 44–59.

Treloar, S A, Heath, A C, Martin, N G (2002), 'Genetic and environmental influences on premenstrual symptoms in an Australian twin sample', *Psychological Medicine* 32 (1), 25–38.

Trivers, R L (1971), 'The evolution of Reciprocal Altruism', *Quarterly Review of Biology* 46, 35–57.

Walter, C, Willett, Meir, J (2002), 'Rebuilding the food Pyramid', *Scientific American*.

Wilkinson, G S (1984), 'Reciprocal food sharing in the vampire bat', *Nature* 308, 181–4.

Wood, W F, Weldon, P J (2002), 'The scent of the reticulated giraffe: *Giraffa camelopardalis reticulata*', *Biochemical Systematics and Ecology* 30, 913–17.

Chapter 8: Genes, Diseases and Getting Older

Andersson, G B (1999), 'Epidemiological features of chronic low-back pain', *Lancet* 354, 581–5.

Barrett, T B, et al. (2003), 'Evidence that a single nucleotide polymorphism in the promoter of the G protein receptor kinase 3 gene is associated with bipolar disorder', *Molecular Psychiatry* 8, 546–57.

Berard, A, et al. (2002), 'Risk factors for the first-time development of venous ulcers of the lower limbs: the influence of heredity and physical activity', *Angiology* 53, 647–57.

Boyd, N F, et al. (2002), 'Heritability of mammographic density, a risk factor for breast cancer', *New England Journal of Medicine* 347, 886–94.

Carroll, D, et al. (2002), 'Admissions for myocardial infarction and World Cup football: database survey', *British Medical Journal* 325, 1439–42.

Clark, C M, Karlawish, J H (2003), 'Alzheimer disease: current concepts and emerging diagnostic and therapeutic strategies', *Annals of Internal Medicine* 138, 400–10.

Fu, Q, et al. (2002), 'A twin study of genetic and environmental influences on suicidality in men', *Psychological Medicine* 32, 11–24.

Harpending, H, Cochran, G (2002), 'In our genes', *Proceeding of the National Academy of Science USA* 99, 10–12.

Hartvigsen, J, et al. (2003) 'Ambiguous relation between physical workload and low back pain: a twin control study', *Journal of Occupational and Environmental Medicine* 60, 109–14.

Haussman, M F, et al. (2003), 'Telomeres shorten more slowly in long-lived birds and mammals than in short-lived ones', *Proceedings of the Royal Society of London B* 2003; 270, 1387.

Hekimi, S, Guarente, L (2003), 'Genetics and the specificity of the aging process', *Science* 299, 1351–4.

Hunter, D J, et al. (2001), 'Genetic contribution to bone metabolism, calcium excretion, and vitamin D and parathyroid hormone regulation', *Journal of Bone Mineral Research* 16, 371–76.

Hunter, D J, et al. (2001), 'Genetic variation in bone mineral density and calcaneal ultrasound: a study of the influence of menopause using female twins', *Osteoporos Int* 12, 406–11.

Indo, Y, et al. (1996), 'Mutations in the TRKA/NGF receptor gene in patients with congenital insensitivity to pain with anhidrosis', *Nature Genetics* 13, 485–8.

Le Flem, L, et al. (2001), 'Thrombomodulin promoter mutations, venous thrombosis, and varicose veins', *Arteriosclerosis Thrombosis and Vascular Biology* 21, 445–51.

Mayeux, R, et al. (1998), 'Utility of the apolipoprotein E genotype in the diagnosis of Alzheimer's disease. Alzheimer's Disease Centers Consortium on Apolipoprotein E and Alzheimer's Disease', *New England Journal of Medicine* 338, 506–11.

McLoughlin, G (2002), 'Is depression normal in human beings? A critique of the evolutionary perspective', *International Journal of Mental Health Nursing* 11, 170–73.

Mead, S, et al. (2003), 'Balancing Selection at the Prion Protein Gene Consistent with Prehistoric Kurulike Epidemics', *Science*.

Mitsos, L M, et al. (2003), 'Susceptibility to tuberculosis: A locus on mouse chromosome 19 (Trl-4) regulates Mycobacterium tuberculosis replication in the lungs', *Proceeding of the National Academy of Science USA*.

O'Brien, J M (2000), 'Environmental and heritable factors in the causation of cancer: analyses of cohorts of twins from Sweden, Denmark, and Finland, by P Lichtenstein, N V Holm, P K Verkasalo, A Iliadou, J Kaprio, M Koskenvuo, E Pukkala, A Skytthe, and K Hemminki', *New England Journal of Medicine* 343, 78–84.

Pericak-Vance, M (2002), 'The genetics of autistic disorder', in Plomin, R, Defries, J, Craig, I W, McGuffin, P (eds), *Behavioural Genetics in the Postgenomic Era* (American Psychological Association), 267–8.

Peto, J, et al. (1999), 'Prevalence of BRCA1 and BRCA2 gene mutations in patients with early-onset breast cancer', *Journal of Natural Cancer Institute* 91, 943–9.

Ralston, S H (2002), 'Genetic control of susceptibility to

osteoporosis', *Journal of Clinical Endocrinology and Metabolism* 87, 2460–6.

Rothstein, M A (2003), Pharmacogenomics: *Social, Ethical, and Clinical Dimensions*, ed. Wiley-Liss, 2003.

Samaras, T T, Elrick, H, Storms, L H (2003), 'Is height related to longevity?', *Life Sci* 72, 1781–1802.

Sambrook, P N, MacGregor, A J, Spector, T D (1999), 'Genetic influences on cervical and lumbar disc degeneration: a magnetic resonance imaging study in twins', *Arthritis Rheum* 42, 366–72.

Shatzky, S, et al. (2000), 'Congenital insensitivity to pain with anhidrosis (CIPA) in Israeli-Bedouins: genetic heterogeneity, novel mutations in the TRKA/NGF receptor gene, clinical findings, and results of nerve conduction studies', *American Journal of Medical Genetics* 92, 353–60.

Shifman, S et al. (2002), 'A highly significant association between a COMT haplotype and schizophrenia', *America Journal of Human Genetics* 71, 1296–1302.

Slagboom, P E, Droog, S, Boomsma, D I (1994), 'Genetic determination of telomere size in humans: a twin study of three age groups', *America Journal of Human Genetics* 55, 876–82.

Traverso, G, et al. (2002), 'Detection of APC mutations in fecal DNA from patients with colorectal tumors', *New England Journal of Medicine* 346, 311–20.

Treloar, S A, et al. (1999), 'Genetic influences on post-natal depressive symptoms: findings from an Australian twin sample', *Journal of Psychological Medicine* 29, 645–54.

Ueda, H, et al. (2003), 'Association of the T-cell regulatory gene CTLA4 with susceptibility to autoimmune disease', *Nature* 423, 506–11.

Westendorp, R G, Kirkwood, T B (1998), 'Human longevity at the cost of reproductive success', *Nature* 396, 743–6.

Williams, J (2002), 'Dementia and Genetics', in Plomin, R, DeFries, J C, Craig, I, McGuffin P (eds), *Behavioural Genetics in the Postgenomic Era* (American Psychological Association).

Zdravkovic, S, et al. (2002), 'Heritability of death from coronary heart disease: a 36-year follow-up of 20 966 Swedish twins', *Journal of Internal Medicine* 252, 247–54.

Zubieta, J K, et al. (2003), 'COMT val158met genotype affects mu-opioid neurotransmitter responses to a pain stressor', *Science* 299, 1240–3.

Chapter 9: Beliefs, Morals and the Afterlife

Alsobrook, J P, Pauls, D L (2000), 'Genetics and violence', *Child and Adolescent Psychiatric Clinics of North America* 9, 765–76.

Bouchard, T J, Jr, McGue, M, Lykken, D, Tellegen A (1999), 'Intrinsic and extrinsic religiousness: genetic and environmental influences and personality correlates', *Twin Research* 2, 88–98.

Brunner, H G, et al. (1993), 'Abnormal behavior associated with a point mutation in the structural gene for monoamine oxidase A', *Science* 262, 578–80.

Caspi, A, et al. (2002), 'Role of genotype in the cycle of violence in maltreated children', *Science* 297, 851–4.

Choi, C, (2002), 'Brain tumour causes incontrollable paedophilia', *New Scientist*, reporting Swerdlow, R, et al. (21 October).

Coid, J, et al. (2001), 'Relation between childhood sexual and physical abuse and risk of revictimisation in women: a cross-sectional survey', *Lancet* 358, 450–4.

Cosmides, I, Tooby, J, Kurzban, R, (2003), 'Perceptions of race', *Trends in Cognitive Science* 7, 173–9.

Davidson, R J, Putnam, K M, Larson, C L (2000), 'Dysfunction in the neural circuitry of emotion regulation – a possible prelude to violence', *Science* 289, 591–4.

De Luca, M, et al. (2003), 'Dopa decarboxylase (Ddc) afects variation in drosophila longevity', *Nature Genetics* 34, 429–433.

Dunbar, R (2003), 'What's God got to do with it?', *New Scientist* (14 June), 38–9.

Eaves, L, et al. (1999), 'Comparing the biological and cultural inheritance of personality and social attitudes in the Virginia 30,000 study of twins and their relatives', *Twin Research* 2, 62–80.

Feingold, L M (1984), *Genetic and Enviromental Determinants of Social Attitudes – PhD thesis* (University of Oxford).

Gwen, K P, Lim, L E, Woo, M (2002), 'The sexual profile of rapists in Singapore', *Journal of Medical Sciences and Law* 42, 51–7.

Kendler, K S, et al. (2003), 'Dimensions of religiosity and their relationship to lifetime psychiatric and substance use disorders', *American Journal of Psychiatry* 160, 496–503.

Kirk, K M, Martin, N G (1999), 'Religion, Values and Health: Unravelling the roles of genes and environment', *Twin Research* 2, 59–179.

Martin, N G, et al. (1986), 'Transmission of social attitudes', *Proceedings of the National Academy of Science USA* 83, 4364–8.

Mason, B (2003), 'Women prefer wimps', *New Scientist*, 2 August, p. 16.

Merrill, L L, et al. (1999), 'Childhood abuse and sexual revictimization in a female Navy recruit sample', *Journal of Traumatic Stress* 12, 211–225.

Merrill, L L, Thomsen, C J, Gold, S R, Milner, J S (2001), 'Childhood abuse and premilitary sexual assault in male Navy recruits', *Journal of Consulting and Clinical Psychology* 69, 252–61.

Miles, D R, Carey, G (1997), 'Genetic and environmental architecture of human aggression', *Journal of Personality and Social Psychology* 72, 207–17.

Moffitt, T E, et al. (1998), 'Whole blood serotonin relates to violence in an epidemiological study', *Biological Psychiatry* 43, 446–57.

Roes, F L, Raymond, M (2003), 'Belief in moralizing gods', *Evolution and Human Behaviour* 24, 126–35.

Thornhill, N W, Thornhill, R (1991), 'An evolutionary analysis of psychological pain following human (*Homo sapiens*)

rape: IV. The effect of the nature of the sexual assault',
Journal of Comparative Psychology 1991; 105, 243–52.
Walker M (2001), 'Insult to injury', *New Scientist* (20 June).

Chapter 10: Conclusion
Brooks, R, Endler, J A (2001), 'Female guppies agree to differ:
phenotypic and genetic variation in mate-choice behavior
and the consequences for sexual selection', *International
Journal of Organic Evolution* 55, 1644–55.
Mischel, W, Shoda, Y, Rodriguez, M I (1989), 'Delay of
gratification in children', *Science* 244, 933–8.
Pedersen, N L, et al. (1989), 'Individual differences in locus of
control during the second half of the life span for identical
and fraternal twins reared apart and reared together',
Journal of Gerontology 44, 100–5.
Vorel, S R, et al. (2002), 'Dopamine D3 receptor antagonism
inhibits cocaine-seeking and cocaine-enhanced brain
reward in rats', *Journal Neuroscience* 22, 9595–603.